Combined Wave And Ray Based Room Acoustic Simulations Of Small Rooms

Challenges and limitations on the way to realistic simulation results

Von der Fakultät für Elektrotechnik und Informationstechnik der Rheinisch-Westfälischen Technischen Hochschule Aachen zur Erlangung des akademischen Grades eines

DOKTORS DER INGENIEURWISSENSCHAFTEN

genehmigte Dissertation

vorgelegt von

Dipl.-Ing. Dipl.-Wirt.-Ing.
Marc Aretz
aus Aachen

Berichter:

Universitätsprofessor Dr. rer. nat. Michael Vorländer
Universitätsprofessor Dr.-Ing. Otto von Estorff

Tag der mündlichen Prüfung: 25. September 2012

Marc Aretz

Combined Wave And Ray Based Room Acoustic Simulations Of Small Rooms

Logos Verlag Berlin GmbH

Aachener Beiträge zur Technischen Akustik

Editor:
Prof. Dr. rer. nat. Michael Vorländer
Institute of Technical Acoustics
RWTH Aachen University
52056 Aachen
www.akustik.rwth-aachen.de

Bibliographic information published by the Deutsche Nationalbibliothek

The Deutsche Nationalbibliothek lists this publication in the Deutsche Nationalbibliografie; detailed bibliographic data are available in the Internet at http://dnb.d-nb.de .

D 82 (Diss. RWTH Aachen University, 2012)

ISBN 978-3-8325-3242-0
ISSN 1866-3052
Vol. 12

Logos Verlag Berlin GmbH
Comeniushof, Gubener Str. 47,
D-10243 Berlin
Tel.: +49 (0)30 / 42 85 10 90
Fax: +49 (0)30 / 42 85 10 92
http://www.logos-verlag.de

Abstract

Classical room acoustic simulation methods based on the principles of geometrical acoustics (GA) have nowadays become an accepted and highly developed tool for acoustic practitioners and researchers in predicting the acoustic characteristics of large rooms like concert halls, theatres or open-space offices. However, when it comes to small rooms, even the most advanced geometrically based methods appear to be flawed due to the inherent negligence of important low frequency wave effects, such as standing waves, diffraction and interference. In order to overcome this limitation the present thesis investigates the potential benefits of the application of the Finite Element Method (FEM) to the modally dominated part of the sound field. However, despite the fact that the FEM is a well-established tool in engineering sciences, which fully captures all relevant wave effects, its application to room acoustics introduces far-reaching and yet unresolved questions regarding the realistic source, boundary and receiver representation.

The present thesis therefore establishes a complete framework for the combination of FE- and GA-based room acoustic simulation results and discusses the inherent challenges and limitations including all aspects of sound generation, sound reflection and sound reception. Moreover, the thesis establishes detailed guidelines for the best-possible determination of all necessary input data for both simulation domains.

The investigations conducted in the course of this thesis aim at two different goals. On the one hand the thesis investigates the influence of selected isolated aspects regarding their influence on the simulation accuracy. In particular, these topics include a study on the potential of the image source method to predict the modal characteristics of the low frequency Room Transfer Function (RTF), a study on the efficient modelling of porous absorbers in the FE domain and finally a study on the possible low frequency coupling of the excitation velocity of a loudspeaker to the sound field at the loudspeaker membrane.

On the other hand the thesis investigates the overall potential of the presented combined approach by conducting extensive objective and subjective comparisons of measurement and simulation results for three types of acoustically relevant small spaces (a scale-model reverberation room, a recording studio and two different car passenger compartments). For each room considerable efforts have been made to obtain a best-possible a-priori assessment of all necessary material and source data for the simulations. However, especially with regard to the determination of the acoustic surface impedances at the room boundaries certain inevitable inaccuracies have to be accepted.

While the presented results reveal an overall good agreement regarding the energy distribution in time and frequency domain for all considered rooms, the results clearly show that as expected the simulation accuracy considerably degrades with increasing complexity of the room geometry and boundary conditions. Moreover, it is important to mention that even with the FEM a precise prediction of the fine structure of the RTF appears impossible in the frequency range far above the Schroeder frequency. It can thus be concluded, that possible fields of application of the FE extension in room acoustic simulations lie in the prediction of the modally dominated low frequency part of the RTF of well defined rooms and in the prediction of sound fields that are strongly affected by near-field or diffraction effects as in the car passenger compartment.

Thus, despite the general potential of the low frequency FE extension to realistically predict the modal structure of the low frequency part of the RTF, the Achilles' heel of room acoustic FE simulations appears to be the determination of realistic impedance conditions on the room boundaries. Consequently the application of the finite element method to room acoustic applications calls for improved measurement techniques for the acoustic surface impedance.

Contents

1. Introduction

The challenges of the realistic simulation of sound fields in enclosed spaces have propelled the efforts of many room acousticians for more than the last 50 years now. Even today, the full simulation and auralization of the sound field in an arbitrarily shaped room including all aspects of sound generation, room transfer paths, sound reception and also sound reproduction still remains an extremely difficult task encompassing various fields of acoustics. These fields span from the fundamentals of sound propagation in enclosed spaces and the deduced simulation algorithms to the determination of suitable source, boundary and receiver characteristics and finally also to psychoacoustic aspects related to the identification of the major perceptual characteristics of sound fields in rooms. The present thesis aims at giving a comprehensive insight into all these aspects of room acoustic simulations. In particular, special focus will be given to the challenges of room acoustic FE simulations which offer a valuable low frequency extension to classical ray based simulations in small rooms. The following subsections give a short review of the relevant fields of acoustics and an outline of the present thesis.

1.1. Short survey on room acoustic simulation history until today

Since the late 1950's many contributions and publications have dealt with the challenge of generating high quality artificial reverberation. While detailed theoretical considerations about the properties of a room impulse response in time and frequency domain were already derived by e.g. Bolt [1947], Bolt and Roop [1950], Schroeder [1954a,b] or Kuttruff and Thiele [1954], early work on the generation of artificial reverberation was at that time constrained to the use of rather simple analog feedback loops. These early attempts were merely concerned with the challenge of producing "natural sounding artificial reverberation" [Schroeder, 1961] rather than replicating the specific room acoustics of a complex shaped, reverberant room.

The first approaches to the latter challenge emerged in the 1960's [Schroeder and Atal, 1963] with the advent of computer-aided simulation tools based on geometrical acoustics theory. According to this theory sound propagation is considered in much the same way as light rays are treated in optics. Although the fundamentals of ray acoustics were already known before the 1960's [Eyring, 1930, Cremer and Müller, 1982], their application to the prediction of sound fields for arbitrary room shapes had to wait for appropriate digital computing systems to come. In 1968 Krokstad et al. were the first to use a computer-based ray tracing (RT) technique to study the envelope of early reflected sound energy in time and space taking into account the specific shape of a room and in 1970 Schroeder presented a complete auralization framework for the generation and playback of computer-aided room acoustic simulations based on the ray tracing technique. First computer implementations of the image source (IS) method emerged just a few years later, but were at first generally restricted to rectangular rooms [Berman, 1975, Gibbs and Jones, 1972, Santon, 1976]. Although it was not the first paper on the topic, the emergence of the image source method in room acoustic simulation is often associated with the landmark paper by Allen and Berkley [1979] who showed that for a rigid rectangular room the image source method yields an equivalent result to the solution of the Helmholtz equation in frequency domain and who gave detailed implementation details for their image source model. Finally, in 1984 Borish extended the image source model to arbitrary polyhedra.

Over the past decades, room acoustic applications of the IS and RT method have been continuously improved and refined and many hybrid approaches have emerged to combine the best features of both methods. Today, such hybrid simulation algorithms are implemented in several commercially available room acoustic simulation packages (e.g. ODEON, CATT, EASE[1]) and are often summarized under the term 'hybrid geometrical acoustics (GA) tools'. Thanks to the continuously increasing performance of available computer systems these packages nowadays allow the simulation of sound fields in arbitrarily shaped and highly reverberant large spaces and are thus widely used by acousticians as an accepted design-aid in the planning of concert halls, opera houses and theatres as well as in the layout of PA systems and sound installations in buildings and public areas. Implementation details on typical state-of-the-art hybrid GA methods can for example be found in [Vorländer, 1989, Heinz, 1993, Naylor, 1993, Dalenbäck, 2010]. In recent years even real-time capable simulation algorithms have been realized [Schröder, 2011].

Although such hybrid methods can generally not capture all relevant wave phenomena of a room sound field (diffraction, standing waves, interference), the simplifying statistical assumptions underlying energy-based GA theory can be considered valid in a frequency range with sufficiently high modal density and overlap (cf. section 3.3 for more details). Already in the 1950's Schroeder [1954a] introduced the so-called 'Schroeder frequency' as a rough measure for the transition between the lower frequencies with little modal overlap and the higher frequencies with strong modal overlap. This frequency is often considered as the lower limit for the applicability of geometrical acoustics. For large concert halls the Schroeder frequency is generally in a range between 20 to 50 Hz and consequently the statistical assumptions underlying geometrical acoustics theory hold for almost the whole audible frequency range. However, simulations in small rooms which have a considerably higher Schroeder frequency necessitate different prediction methods to realistically model the modal effects that dominate the sound field in the frequency range below the Schroeder frequency.

While substantial theoretical contributions in small room acoustics have already been made by Bolt [1947] and Cremer and Müller [1982] in the 1940's and 50's the prediction of modally dominated sound fields was, at that time, generally restricted to analytical solutions for rather simple geometries and boundary conditions. Despite the fact that suitable formulations for the numerical solution of the Helmholtz wave equation in complex geometry were already published in the late 1960's [Gladwell, 1966, Schenck, 1968], it was

[1] `www.odeon.dk`, `www.catt.se`, `ease.afmg.eu`

not until the 1990's that the rapidly increasing computational power allowed an efficient application of numerical wave based methods such as the Finite Element (FE) or Boundary Element (BE) method to room acoustic problems. Early work on the coupling of low frequency and high frequency models has been pioneered by Kleiner and Granier [Kleiner et al., 1995, Granier et al., 1996] in the mid-nineties who reported unsatisfactory results from their combination methods. The applied combination techniques appear to have been mainly flawed by the restricted frequency range for the numerical FEM simulation. More recent contributions were for example made by Bansal et al. [2005] and Summers et al. [2004]. Nowadays, it is generally possible to simulate the whole frequency range below the Schroeder frequency using FEM or BEM on a standard PC in reasonable computation times.

Consequently, the extension of established classical geometrical acoustics (GA) tools by wave-based numerical simulations opens the door to the realistic full bandwidth simulation of a whole new range of acoustically interesting small spaces, like reverberation rooms, recording studios or car passenger compartments and first promising results have been published lately [Aretz, 2009, Aretz and Vorländer, 2010, 2009, Pelzer et al., 2011a].

1.2. Boundary conditions for room acoustic simulations

Any kind of room acoustic simulation requires the establishment of a geometric room model and suitable mathematical models for the source radiation, the sound reflections at the room boundaries and the sound reception. Under the assumption of a sufficiently accurate sound propagation model it is the accuracy of these models and their assigned input data that are the leading factors influencing simulation quality and accuracy. Taking into account that contemporary CAD (Computer Aided Design) tools facilitate the design of high quality geometric room models, even more emphasize is put on the need for a realistic representation of the acoustic source, boundary and receiver conditions. In order to thoroughly understand how these factors influence the simulation accuracy a clear cut distinction needs to be drawn between the inherent simplifying assumptions made in these models and the uncertainty in the determination of their required input data. Moreover, it is important to note that the commonly applied source, boundary and receiver models vary considerably depending on the underlying sound field model (e.g. FEM and GA). Figure 1.1 gives a quick overview of the required models for geometrical acoustics and the finite element method. A detailed discussion on this topic will be given in sections 4.1.2, 4.2.4 and in chapters 5 and 6.

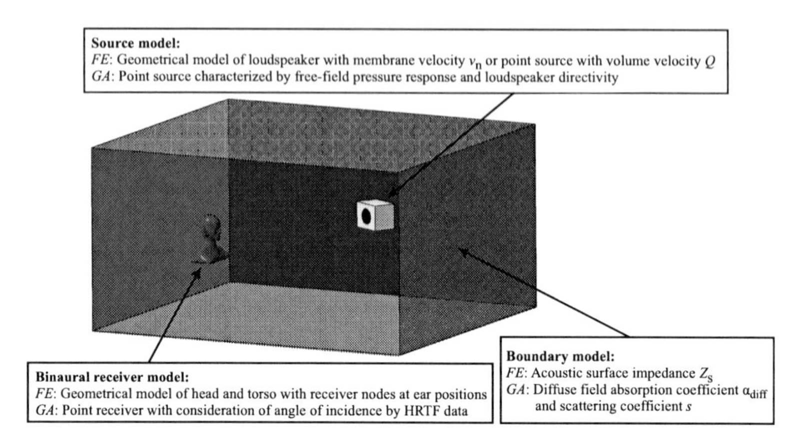

Figure 1.1.: Overview of source, boundary and receiver models in geometrical and wave based room acoustic simulations

1.3. Key characteristics of sound fields in rooms and their relation to human auditory perception

Although the Helmholtz wave equation with a set of suitable boundary conditions (cf. chapter 3.3) can in principle give a very close physical model of the airborne sound propagation in enclosed spaces, the immense complexity of sound fields in real rooms makes it necessary for room acoustic simulation tools to resort to simplifying assumptions regarding the sound propagation model and the source, boundary and receiver characteristics. However, due to the limited spatial, time and frequency resolution of the human auditory system, these simplifications do not necessarily need to compromise the subjectively perceived simulation quality and thus high demands on the simulation accuracy can in principle still be met. Consequently, the aim of a room acoustic simulation is generally not the exact physical reproduction of the sound field in a real room but the synthesis of a similar sound at the listeners ears that is at best perceptually indistinguishable from the real sound. This reasoning however makes the comparison of room impulse responses subject to psychoacoustic evaluation methods and thus makes it difficult to benchmark the quality of room acoustic simulation results. In order to approach this problem it is therefore in a first step crucial to identify and classify the key characteristics of sound fields in rooms that most strongly relate to human auditory perception. Moreover, in a second step, it is important to determine subjective difference limens for the identified measures that have to be met to sufficiently replicate the room acoustic footprint of an existing room including its source and receiver. In this context many attempts have been made over the last decades to define a comprehensive set of perceptually relevant and objectively measurable room acoustic parameters. The essence of these various studies is summarized in the ISO standard “Acoustics - Measurement of room acoustic parameters

- Part 1: Performance spaces (ISO 3382-1:2009)" and many publications have dealt with the determination of suitable difference limens for these parameters (an overview of these so-called "just noticeable differences" is given in the Annex A of the ISO 3382 standard. Important publications on the topic were for example contributed by Seraphim [1958], Cox et al. [1993], Vorländer [1995], Witew et al. [2005]). However, since these parameters can of course not cover all perceptual aspects of a room impulse response (cf. e.g. Soulodre and Bradley [1995]), investigations on the quality of room acoustic simulations still have to resort to subjective evaluations and testing methods.

1.4. Outline of the present thesis

The present thesis applies a combined FE and hybrid GA simulation approach to realistically model the sound field in small rooms for the whole audible frequency range and discusses two extensive application examples, where thorough comparisons between measured and simulated room transfer functions are conducted to demonstrate the potential and possible problems of the presented simulation approach. Figure 1.2 gives an overview of the applied combined simulation algorithm.

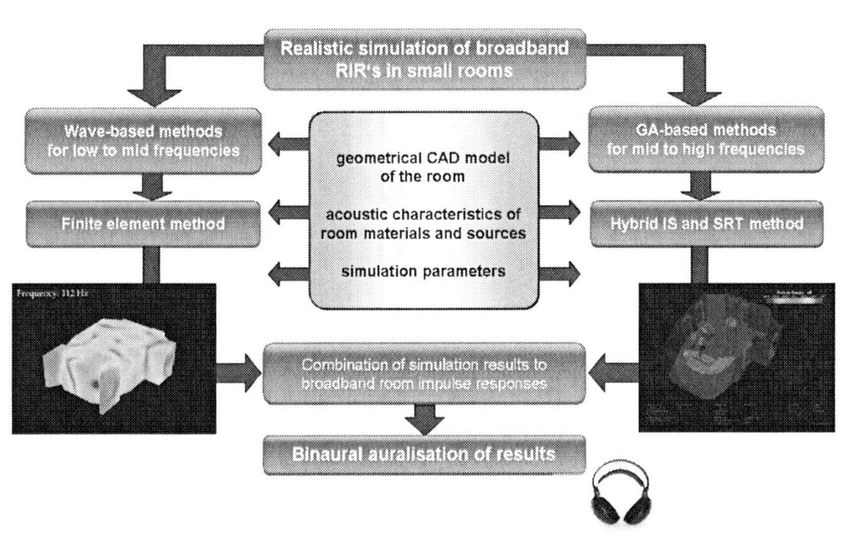

Figure 1.2.: Overview of combined FE and GA simulation approach

Before looking into these application examples, chapters 2, 3 and 4 lie down the theoretical foundations for the applied sound field models and simulation algorithms. In particular, chapter 2 summarizes the necessary signal processing fundamentals, chapter 3 deals with the relevant theory of room acoustics from a wave and ray based point of view and chapter 4 discusses the applied simulation models and algorithms with special focus on the crucial limitations of these models and their required input data. Furthermore chapter 4 presents the combination method that is used to combine the FE and GA results in the frequency domain at a given cross-fade frequency. Next, chapters 5 and 6 take up on the topic of the determination of realistic simulation input data and discuss the limits and problems of the presented measuring and modeling techniques. The according chapters are subdivided into parts dealing with the determination methods for (a) the acoustic boundary conditions, (b) the source conditions and (c) the receiver representation in the simulations. Thus chapters 4 to 6 contain all relevant information on the combined simulation algorithm and the determination methods for all necessary input data. Chapter 7 then presents three preliminary simulation studies, that deal with selected aspects of room acoustic simulations in small rooms; section 7.1 presents a study on the applicability of the image source method in the modally dominated frequency range, section 7.2 discusses the performance of different porous absorber models in room acoustic FE simulations and section 7.3 deals with possible coupling effects between low frequency room modes and the diaphragm velocity of a woofer loudspeaker.

Based on the results of these preliminary studies and on the simulation framework presented in the earlier chapters, chapters 8 and 9 finally present the two application studies which are based on the comparison of measurement and simulation results obtained in a recording studio control room and a car passenger compartment. In both studies the combined simulation approach is used to simulate the room sound field in the whole audible frequency range. Special focus is put on the assessment of the required input data which was determined to the best possible extent in both studies. Finally, chapter 10 concludes the thesis and gives an outlook on possible future work.

2. Signal processing fundamentals

The present chapter gives a very short introduction to the important concepts of Fourier analysis and LTI systems which form the basis for the measurement, processing and evaluation of acoustic signals and particularly sound signals in rooms, which are the focus of this thesis. For the sake of brevity all formulas are only given for signals that are continuous in time and frequency dimension, although contemporary computer-based measurement systems and processing tools of course use discrete time signals and frequency spectra. For an extensive introduction to these topics and a discussion of the peculiarities of discrete signal processing the reader is referred to the relevant standard literature [Ohm and Lüke, 2010, Oppenheim et al., 1997, 1999].

2.1. Fourier transformation

The processing and analysis of acoustic signals is generally conducted in time and frequency domain, where both domains are linked by the Fourier Transformation. The respective transformation integrals are given as follows:

$$\underline{S}(f) = \int_{-\infty}^{\infty} s(t) e^{-j2\pi ft} \, \mathrm{d}t \tag{2.1}$$

$$s(t) = \int_{-\infty}^{\infty} \underline{S}(f) e^{j2\pi ft} \, \mathrm{d}f, \tag{2.2}$$

where f is the frequency and t the time. This means that a time signal $s(t)$ can be represented by the superposition of weighted orthonormal harmonic basis functions $e^{j2\pi ft}$, where the weighting factors $\underline{S}(f)$ are determined by the Fourier Transformation and are a measure of how much a single harmonic component contributes to the overall time signal. The ensemble of weighting factors $\underline{S}(f)$ thus forms the frequency spectrum of the time signal $s(t)$ and gives an equivalent representation of the signal on the basis of the orthonormal basis functions $e^{j2\pi ft}$.

Based on these time and frequency domain considerations, the transmission path between a given source and receiver in a room can be described by its room impulse response (RIR) $h(t)$ or equivalently by its room transfer function (RTF) $\underline{H}(f)$, which are linked by the Fourier transformation. Although $h(t)$ and $\underline{H}(f)$ are equivalent representations of the same transmission system, both representations allow very different but equally valuable insights into the acoustic characteristics of a room. The impulse response $h(t)$ at a certain receiver position can be interpreted as the response of the room to an ideal dirac impulse which is emitted by a sound source in the room (The Dirac delta function $\delta(x)$, which was introduced by theoretical physicist Paul Dirac, can be interpreted as a

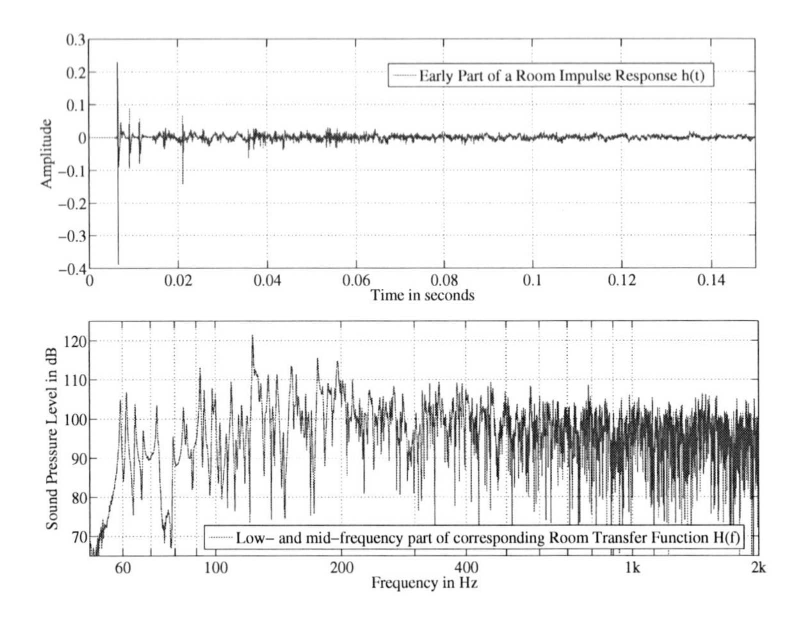

Figure 2.1.: Time and frequency domain representation of the sound transmission path in an exemplary room between a given source and receiver position.

very short impulse with very high amplitude at $x = 0$ and can for example be derived as the limit of a sequence of Gaussians: $\delta_a(x) = \frac{1}{a\pi}e^{-x^2/a^2}$ with $a \to 0$.) Consequently the RIR describes the temporal decay process starting with the arrival of the direct sound impulse which is then followed by many delayed and attenuated impulses reaching the receiver over various wall reflections. On the other hand the RTF $\underline{H}(f)$ of the room can be interpreted from an intuitive point of view as the steady state response of the room to a harmonic excitation with frequency f. Thus, if a room is excited by a sinusoidal signal with a constant amplitude and slowly increasing frequency over time (such that the room reaches its steady state mode for every excited frequency) the room transfer function could be obtained with a simple level recorder. Figure 2.1 shows an example of a room impulse response $h(t)$ with its respective frequency transfer function $\underline{H}(f)$.

2.2. Room transfer path as an LTI system

In a next step, it is interesting to see, how the pressure response $g(t)$ of a room at a given receiver position to an arbitrary excitation signal $s(t)$ can be given as a function of the RIR $h(t)$ and RTF $\underline{H}(f)$ of a room. It is therefore useful to make the reasonable assumption that the sound transmission in a room can be modeled as a so-called LTI (Linear Time-Invariant) system, since in this case the time domain response $g(t)$ of the room to an excitation signal $s(t)$ can be calculated by the convolution integral using the RIR $h(t)$ or equivalently the frequency domain response $\underline{G}(f)$ can be calculated by multiplication of the excitation spectrum $\underline{S}(f)$ with the room transfer function $\underline{H}(f)$:

$$g(t) = \int_{-\infty}^{\infty} s(\tau)h(t-\tau)\,\mathrm{d}\tau = s(t) * h(t), \tag{2.3}$$

$$\underline{G}(f) = \underline{S}(f) \cdot \underline{H}(f). \tag{2.4}$$

Using this system theoretic approach the acoustic transmission path in a room can be described by a black box, which is fully characterized by its source and receiver position dependent room impulse response $h(t)$ and transfer function $H(f)$ (cf. figure 2.2). The following sections will therefore discuss how this black box can be described based on the fundamental equations for wave propagation in fluid media.

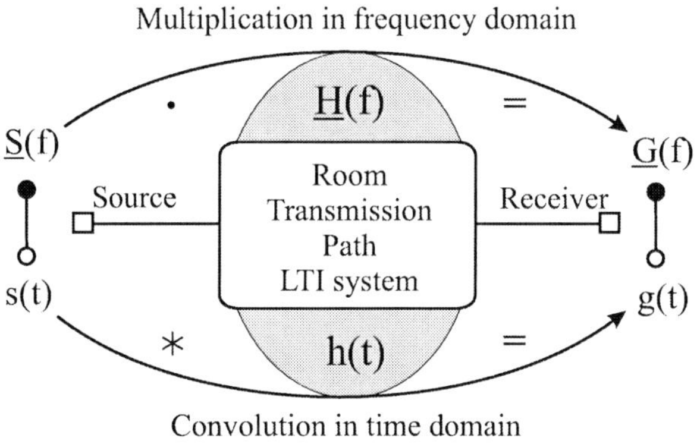

Figure 2.2.: The acoustic transmission path in a room as an LTI system black box (after Vorländer [2007]).

3. Sound fields in rooms - Theoretical foundations

The field of room acoustics is based on the theory of wave propagation of airborne soundfields within three-dimensional enclosures including the consideration of the acoustical reflection properties of the room boundaries. A detailed derivation of the underlying physical and mathematical foundations is given in many text books on acoustics [Kuttruff, 2000, Cremer and Müller, 1982]. The present chapter will therefore only briefly summarize the key concepts of room acoustics which are relevant to the understanding of the geometrical and wave based simulation models used in this thesis. Special focus is given to the theoretical link between wave and ray acoustics and to the inherent limitations of the used sound propagation, boundary and source models.

Taking into account the immense complexity of real sound fields in rooms we will first start by considering the wave propagation in an unbounded space and than discuss how sound waves are reflected at a single unbounded planar boundary surface. On this basis we will then extend our considerations to the case of multiple reflections in an enclosed space, where we approach the problem both from a wave theoretical point of view based on the solution of the Helmholtz equation as well as from a geometrical acoustics point of view, where the sound propagation is modeled with the concept of sound rays.

3.1. Sound propagation in fluid media

In airborne acoustics the sound propagation in fluid media (in the free field or an enclosed space) is most commonly described in the frequency domain by the Helmholtz wave equation. This equation forms the basis of all linear sound propagation models and describes the harmonic pressure, density and temperature oscillations of a fluid particle about its mean position in a wave field. The Helmholtz wave equation is derived on the basis of the fundamental equations for (a) the conservation of momentum (2nd Newtonian equation of motion) and (b) the conservation of mass (equation of continuity in hydrodynamics), which are given as follows:

$$\text{(a)} \quad \rho_0 \frac{\partial \underline{\boldsymbol{v}}}{\partial t} = -\text{grad}\, \underline{p}, \qquad \text{(b)} \quad \rho_0 \,\text{div}\, \underline{\boldsymbol{v}} = -\frac{\partial \underline{\rho}}{\partial t}, \tag{3.1}$$

where $\underline{p}$ and $\underline{\rho}$ are the complex valued pressure and density fluctuations around their respective static values p_0 and ρ_0, $\underline{\boldsymbol{v}}$ is the particle velocity vector and t the time. Further assumptions for the derivation of the Helmholtz equation are that (a) the pressure and density fluctuations are small compared to the static values p_0 and ρ_0, (b) the sound velocity $\underline{\boldsymbol{v}}$ is small compared the speed of sound $c = 344\,\mathrm{m/s}$ and (c) the fluid can be considered as an ideal gas with adiabatic temperature variations. These conditions are generally met in good approximation for airborne sound levels within the human audible range. Taking into account that under condition (c) the sound pressure and fluid density fluctuations are related by

$$\underline{p} = c^2 \underline{\rho}, \quad \text{with} \quad c = \sqrt{\kappa \frac{p_0}{\rho_0}}, \tag{3.2}$$

where κ is the adiabatic exponent of air[1], the equations in 3.1 and 3.2 can be merged into the following wave equation by elimination of the particle velocity $\underline{\boldsymbol{v}}$ and the variable part of the density $\underline{\rho}$:

$$\Delta \underline{p} = \frac{1}{c^2} \frac{\partial^2 \underline{p}}{\partial t^2}, \tag{3.3}$$

with the Laplace operator Δ = div grad. In the case of harmonic time signals with $\underline{p}(\boldsymbol{x}, t) = \underline{p}(\boldsymbol{x}) e^{j\omega t}$ the wave equation can be transformed into the Helmholtz equation:

$$\Delta \underline{p} + k^2 \underline{p} = 0, \quad \text{with} \quad k = \frac{\omega}{c} \quad \text{and} \quad \omega = 2\pi f, \tag{3.4}$$

where the constant k is called the propagation constant and ω is called angular frequency. Taking into account that the time dependence of $\underline{p}(\boldsymbol{x}, t)$ was deliberately chosen such that it conforms to the earlier defined basis functions of the Fourier transformation it is clear that the extension to arbitrary time signals can be done by use of the inverse Fourier Transformation (cf. equation 2.2) of the solutions $\underline{p}(\boldsymbol{x})$. A detailed derivation of the Helmholtz equation in fluid acoustics with more information on the underlying equations and assumptions can for example be found in Kuttruff [2000], Morse and Ingard [1986] or Cremer and Müller [1982].

A fundamental solution to the homogeneous Helmholtz equation in free space is given by the plane wave solution

$$\underline{p}(\boldsymbol{x}) = \underline{\hat{p}} e^{-j\boldsymbol{k}\boldsymbol{x}} \tag{3.5}$$

where $\underline{\hat{p}}$ is the amplitude of the pressure wave and $\boldsymbol{k} \cdot \boldsymbol{x}$ is the scalar product of the observation point $\boldsymbol{x}$ in cartesian coordinates and the propagation vector $\boldsymbol{k}$ with $|\boldsymbol{k}| = k$. The plane wave solution thus describes a harmonic sound wave that is spatially invariant on each plane orthogonal to the propagation vector $\boldsymbol{k}$. It is easy to show that equation 3.5 satisfies the homogeneous Helmholtz equation in free space. Although this wave type is a somewhat idealized theoretical construct, since it can only be generated approximately in special experimental setups, it is the basis of many theoretical considerations regarding the reflection, scattering or diffraction of sound waves. In particular it is often used to approximate narrow sections of spherical waves (see below) at sufficient distance from an arbitrary source. As we will see in the next section, this admissible simplification considerably facilitates the physical description of the reflection of sound waves at a planar boundary.

[1] In literature κ is more often referred to as the isentropic exponent.

If sound sources are present inside the considered volume, the Helmholtz equation needs to be extended to its inhomogeneous form:

$$\Delta \underline{p} + k^2 \underline{p} = -j\omega\rho_0 \, \underline{q}(\boldsymbol{x}), \tag{3.6}$$

where $\underline{q}(\boldsymbol{x})$ is the density function of a continuously distributed volume velocity $\underline{q}(\boldsymbol{x})\,\mathrm{d}V$. The right hand side of this equation is called the source function of the partial differential equation [Ehlotzky, 2007]. From a physical point of view this term originates from adding the source term $\rho_0 \, \underline{q}(\boldsymbol{x})$ (which accounts for the volume displacement generated by the distributed sound source) to the conservation of mass equation (cf. eq.: 3.1).

A fundamental solution to the inhomogeneous Helmholtz equation can be given analytically in the case of a point source with a volume velocity $\underline{Q}$, which is represented by a spatial Dirac impulse on the right hand side of equation 3.6. In this case the resulting time harmonic pressure function is given by

$$\underline{p}(r) = \frac{j\omega\rho_0 \underline{Q}}{4\pi r} e^{-jkr}. \tag{3.7}$$

Thus, the solution constitutes a spherical wave of frequency f, where r is the distance of the observation point to the position of the sound source. Assuming further that the spectrum of the volume velocity $Q(f)$ is proportional to f^{-1}, the point source generates a flat pressure frequency spectrum with amplitude proportional to r^{-1} and phase equal to $\arg(\underline{p}) = -jkr$. It can be shown [Ohm and Lüke, 2010] that this spectrum corresponds to a dirac impulse in time domain which is delayed by $\frac{r}{c}$ seconds and where the amplitude decreases with $\frac{1}{r}$. This consideration is especially important when discussing the reflection of sound waves at planar surfaces by use of the image source method which is discussed in section 3.2.3. Although such a point source is, much like the above mentioned plane wave, a theoretical construct (an infinitesimally small volume with finite volume displacement), which cannot be realized physically, it has to be mentioned that at sufficient distance many real sound sources of finite size can be modeled by the aggregation of one or several point sources (multipoles).

Figure 3.1 illustrates the plane and spherical wave solutions as a snap shot where the dotted black lines indicate wavefronts with identical phase. In this figure it was arbitrarily chosen that the plane wave propagates in the x-direction and the spherical wave originates at $\boldsymbol{x} = \boldsymbol{0}$. The distance between the wavefronts is given by the so-called wavelength $\lambda = \frac{c}{f}$.

3.2. Sound reflection at an extended planar boundary

It was already mentioned in the introductory part of this thesis that the specification and parameterisation of suitable sound reflection models, which describe the physical phenomena related to the reflection of a sound wave from a boundary surface in a room, constitute major challenges in the prediction of room acoustic sound fields. While chapter 5 will deal with the metrological uncertainty in the parameterisation of the boundary models which are commonly used in room acoustics, the present section will focus on the used boundary models themselves and their inherent limitations. Section 3.2.1 therefore introduces the

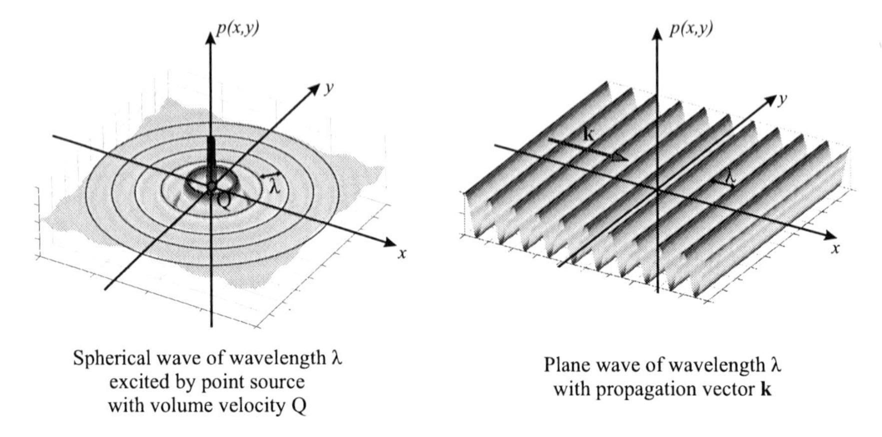

Figure 3.1.: Snap shot of a spherical and plane wave of wavelength λ at a time instant t. Dotted black lines represent regions with identical phase and are called wave fronts. The z-axis represents the sound pressure value.

complex acoustic surface impedance $\underline{Z}_S$ and the concept of 'locally reacting' boundary surfaces. Next, sections 3.2.2 and 3.2.3 discuss how the reflection of plane and spherical waves from an extended planar surface can be described as a function of the acoustic surface impedance $\underline{Z}_S$ and introduce the plane wave reflection factor $\underline{R}$ and the absorption coefficient α. The exact solutions to these elementary reflection scenarios contain fundamental implications regarding the simplifications of the boundary models commonly used in room acoustic simulations. An extensive elaboration on the presented topic is given in the books on sound absorption by Mechel [1989, 1995, 1998]. More detailed references to specific sections in Mechel's books will be given in the following subsections wherever this is considered helpful.

3.2.1. Acoustic surface impedance

The present section discusses how the physical behaviour of a boundary surface that is hit by an incident sound wave can be described by suitable boundary conditions for the Helmholtz equation. These acoustic boundaries can generally be described as layered configurations of porous materials, elastic plates or solids, foils, membranes or perforated plates. Thus a full model of the sound reflection at such a boundary would necessitate the calculation of a mutually coupled model, with suitable sound propagation models for all materials involved in the layered boundary configuration. In room acoustic modeling this is however not desired since the model generally only extends over the fluid domain (i.e. the air inside the room) in order to keep the complexity at a manageable level. In the following it will therefore be shown under which conditions the physical behaviour of a boundary wall can be described independent of the incident sound field by its acoustic surface impedance $\underline{Z}_S$, which is defined as the ratio of sound pressure $\underline{p}$ and surface normal velocity $\underline{v}_n$ at a room boundary.

In order to better understand the circumstances under which the acoustic surface impedance sufficiently describes the physical behaviour of a boundary wall, the problem shall be approached from a more physical point of view. Following Mechel [1989, p.23ff] a general type of boundary condition at the boundary of a nonviscous fluid media can be formulated by demanding that the sound pressure and the surface normal velocity on each side of the boundary surface have to be continuous. It can be shown [Mechel, 1989, p.24] that this statement is equivalent to the postulation of identical surface impedances $\underline{Z}_{\mathrm{S}} = \frac{\underline{p}}{\underline{v}_n}$ at the boundary interface and identical tangential components of the propagation constants in both adjoining media. These equivalent formulations are however rather 'coupling conditions' than 'boundary conditions', since they are not independent of the type of the sound field in front of the boundary. However, under the assumption that sound propagation behind the boundary is only possible in the perpendicular direction to the boundary surface, the behaviour of an acoustic boundary can be described solely by its acoustic surface impedance $\underline{Z}_{\mathrm{S}}$, which is in this case independent of the incident sound field. Such boundaries are called 'locally reacting' [Mechel, 1989, p.23], since the surface normal velocity $\underline{v}_{\mathrm{n}}$ at a point $\boldsymbol{x}$ at the boundary depends only on the sound pressure $\underline{p}$ at the exact same point $\boldsymbol{x}$ and is independent of the surrounding pressure distribution [Kuttruff, 2000, p.30]. In other words this means that there exists no coupling between laterally adjacent points in the boundary wall for a 'locally reacting' material. This is obviously not true for elastic plates since adjacent elements are coupled by their bending stiffness. Moreover, in the case of porous absorber materials this approximation is only admissible if the absolute value of the propagation constant of the porous absorber is much higher than that of air (cf. section 7.2.3). However, for the moment we only want to summarize the following important statement:

> *Under the assumption that a room boundary is locally reacting its acoustic reflection characteristics can be fully described by its acoustic surface impedance $\underline{Z}_{\mathrm{S}}$, which is independent of the incident sound field and which can be specified as a boundary condition for the Helmholtz equation in a bounded space.*

3.2.2. Plane wave reflection at an extended planar boundary

We will first consider a plane wave that is incident at an angle θ upon an infinite planar boundary surface. Under the assumption that the boundary surface is homogeneous along its lateral dimension the resulting sound field in front of the boundary can be described as the superposition of an incident and a specularly reflected plane wave. This is shown in figure 3.2 (a). If we assume that part of the incident sound wave gets absorbed at the boundary surface, where the term 'absorbed' includes both sound transmission and transformation into other energy forms (like heat) in the boundary layer, the amplitude reduction and phase shift of the reflected wave with respect to the incoming wave can be described by the so-called plane wave reflection factor $\underline{R} = |\underline{R}| \exp(j\gamma)$. Thus the pressure $\underline{p}$ and the surface normal velocity $\underline{v}_{\mathrm{n}}$ at the boundary surface at $z = 0$ can be written as:

$$\underline{p}(0,t) = \underline{\hat{p}}(1+\underline{R})e^{j2\pi ft} \quad \text{and} \quad \underline{v}_{\mathrm{n}}(0,t) = \frac{\underline{\hat{p}}\cos(\theta)}{\rho_0 c}(1-\underline{R})e^{j2\pi ft}, \tag{3.8}$$

where $\underline{\hat{p}}$ is the amplitude of the incident sound wave. The surface normal velocity was calculated from the incoming and reflected pressure waves by the equation for the conservation of momentum 3.1. By subdividing the sound pressure $\underline{p}$ and the surface normal velocity $\underline{v}_{\mathrm{n}}$ at $z = 0$ it is possible to relate the plane wave reflection factor $\underline{R}(\theta)$ to the acoustic surface impedance $\underline{Z}_{\mathrm{S}}$ by

$$\underline{Z}_{\mathrm{S}} = \frac{\underline{p}}{\underline{v}_{\mathrm{n}}}\Big|_{z=0} = \frac{Z_0}{\cos\theta}\frac{1+\underline{R}(\theta)}{1-\underline{R}(\theta)} \quad \Leftrightarrow \quad \underline{R}(\theta) = \frac{\underline{Z}_{\mathrm{S}} - {}^{Z_0}/_{\cos\theta}}{\underline{Z}_{\mathrm{S}} + {}^{Z_0}/_{\cos\theta}}, \tag{3.9}$$

where $Z_0 = \rho_0 c$ is called the characteristic impedance of air. From the second equation it can be seen, that even in the case of a 'locally reacting' boundary with angle independent impedance $\underline{Z}_{\mathrm{S}}$ the reflection factor $\underline{R}$ depends on the angle of incidence θ. Another important conclusion from this equation is that in the case of plane wave incidence on a locally reacting boundary the sound field can be fully described by deducing the reflection factor $\underline{R}(\theta)$ for the considered angle of incidence from the acoustic surface impedance $\underline{Z}_{\mathrm{S}}$.

Another important quantity for the characterization of room acoustic boundaries is the absorption coefficient α which is defined as the ratio of the absorbed sound power Π_{a} to the incident sound power Π_{i}, when a plane wave is incident on a flat extended boundary surface. The absorption coefficient for a given angle of incidence θ can be given as a function of the plane wave reflection factor $R(\theta)$ by:

$$\alpha(\theta) = \frac{\Pi_{\mathrm{a}}}{\Pi_{\mathrm{i}}} = 1 - |\underline{R}(\theta)|^2. \tag{3.10}$$

In the idealized case of diffuse sound incidence on an absorber surface, which means that the sound intensity is homogeneously distributed over all angles of incidence with $I(\theta,\phi) = const$, it is further possible to define the so-called diffuse field absorption coefficient α_{diff}. In this case the total incident and absorbed sound power can be expressed as a superposition of the contributions from homogeneously distributed plane sound waves that hit the absorber surface at different angles of incidence θ. In mathematical notation this can be expressed by integrating all intensity contributions over a unity half sphere, which results in the so-called Paris formula [Paris, 1927] (cf. also Kuttruff [2000]):

$$\alpha_{\mathsf{diff}} = \frac{\Pi_{\mathrm{a}}}{\Pi_{\mathrm{i}}} = \frac{\int_0^{\frac{\pi}{2}} \alpha(\theta) I \sin(\theta)\cos(\theta)\,\mathrm{d}\theta}{\int_0^{\frac{\pi}{2}} I \sin(\theta)\cos(\theta)\,\mathrm{d}\theta} = 2\int_0^{\frac{\pi}{2}} \alpha(\theta)\sin(\theta)\cos(\theta) d\theta \tag{3.11}$$

It is important to note that the cosine term in the integral stems from the fact that the integration only extends over the surface normal fractions of sound intensity $I_{\mathrm{n}} = I\cos(\theta)$, while the sinus term stems from the integration over all solid angles $\mathrm{d}\Omega = \sin\theta\,\mathrm{d}\theta\,\mathrm{d}\phi$ on the unity half sphere. It will be discussed in sections 7.1 and 7.2 that from an empirical point of view an upper frequency bound of 78° is often considered more appropriate since ideally diffuse incidence conditions can hardly be established in real rooms and the corrected upper frequency bound of 78° often shows a better fit with measured data (cf. fn. 2 in section 7.1.4). In order to get a clear cut distinction the 90° case is therefore denoted as "diffuse incidence" and the 78° case as "field incidence" respectively.

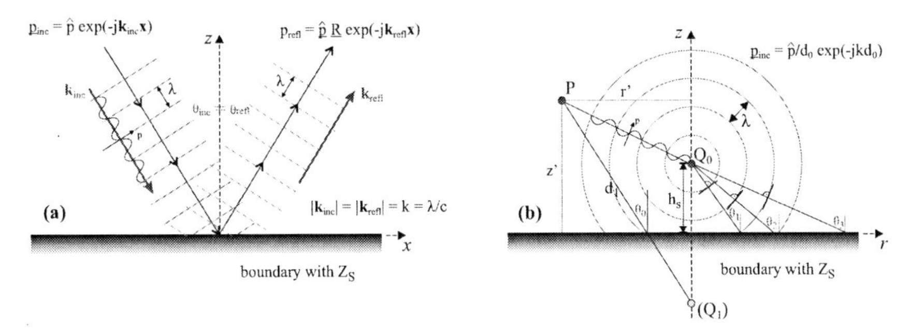

Figure 3.2.: Schematic sketch with coordinate definitions for the reflection of a plane and spherical wave at an extended planar boundary with surface impedance Z_S. Thin blue and green lines indicate wave fronts.

3.2.3. Point source above an extended planar boundary

In the case of spherical sound incidence on an extended planar boundary, things become much more difficult. This is due to the fact that for a point source Q_0 at distance h_s to the absorber plane the angle of incidence on the absorber plane gradually decreases from perpendicular to grazing the further one gets away from the source. Consequently, different portions of the incident sound wave are reflected differently depending on their angle of incidence on the absorber. This is indicated in figure 3.2 (b). In order to give an exact solution for this setup the incident spherical sound wave can be decomposed into an infinite sum of plane waves [Mechel, 1989, p.516ff]. The reflected wave is then calculated by the superposition of the reflected plane waves where the summation has to account for the angle dependence of the reflection factor and the phase shift on the way from the source point over the reflection point to the receiver point for each plane wave contribution. Using this approach the resulting sound field for a normalized[2] point source on the z-axis at distance h_s to the absorber plane which lies perpendicular to the z-axis can thus be given in cylinder coordinates (r, z) as follows [Mechel, 1989, p.521ff], [Suh and Nelson, 1999]:

$$\begin{aligned} \underline{p}_{\text{total}}(r,z) &= \underline{p}_{\text{inc}} + \underline{p}_{\text{refl}} \\ &= \frac{1}{|\boldsymbol{d}_0|} e^{-jk|\boldsymbol{d}_0|} + jk \int_{\Gamma_\theta} J_0(kr' \sin(\underline{\theta})) e^{-jk(z'+h_s)\cos\underline{\theta}} \underline{R}(\theta) \sin\underline{\theta} \, \mathrm{d}\underline{\theta} \ , \quad (3.12) \end{aligned}$$

where $\boldsymbol{d}_0$ is the distance vector between the receiver point $P = (r', z')$ and the source point $Q_0 = (0, h_s)$ according to figure 3.2 (b), J_0 is the Bessel function of zeroth order and Γ_θ is a suitable integration path for the angles $\underline{\theta}$ in the complex plane. The integration path is chosen such that the integral over the plane wave contributions spans over all possible directions of the propagation vector $\boldsymbol{k} = (k_x, k_y, k_z)$ with $|\boldsymbol{k}| = k$ and $-\infty < k_x, k_y < +\infty$ (the component k_z is then given by $\sqrt{k^2 - k_x^2 - k_y^2}$). It is important to mention that in

[2]'Normalized' means in this case that the volume velocity $Q(\omega)$ is chosen such that the point source has a frequency constant pressure amplitude of 1 at a distance of 1 m.

this case it is not sufficient to confine oneself to real valued angles θ and Mechel [1989, p.518] suggests a piecewise linear integration path through the points $(0,0) \rightarrow (\frac{\pi}{2},0) \rightarrow (\frac{\pi}{2},-\infty)$, where the first vector entry gives the real part of $\underline{\theta}$ and the second entry the imaginary part respectively. Many publications have dealt with the problem of finding a smart integration path Γ_θ in order to give useful approximations to the integral formula in equation 3.12. However, this shall not be a concern of this thesis and the reader is referred to the excellent elaboration and literature review on the topic by Mechel [1989, p.522ff]. More importantly it has to be mentioned that a simple approximation to the exact formulation in equation 3.12 can be given by modeling the reflected part of the sound field by a mirror source at $z = -h_s$ as follows:

$$\underline{p}^*_{\text{refl}} = \underline{R}(\theta_0)\frac{1}{|\boldsymbol{d}_1|}e^{-jk|\boldsymbol{d}_1|}, \tag{3.13}$$

where $\boldsymbol{d}_1$ is the distance vector between the receiver point $P = (r', z')$ and the mirror source point $Q_1 = (0, -h_s)$ and θ_0 is the angle between the boundary surface normal and $\boldsymbol{d}_1$ which we call "specular reflection angle" (cf. fig. 3.2 (b)). In this case the reflected wave corresponds to a spherical wave with a directivity function $\underline{R}(\theta)$, which accounts for the angle dependency of the reflection coefficient. Although this model satisfies the boundary conditions everywhere on the planar boundary by weighting the mirror point source with $\underline{R}(\theta)$, the error of this model lies in the sound field generated by the mirror source, since an infinitesimal point source with directivity factor $\underline{R}(\theta)$ does not satisfy the wave equation [Mechel, 2002]. While this leads to errors in the near field of the source, Mechel [2002] points out that the resulting errors are in most cases acceptable. In particular he states that considerable errors are only expected if (a) the sum of heights of Q_0 and P over the wall is not large compared to the considered wavelengths, (b) the specular reflection angle θ_0 gets close to grazing incidence or (c) if the reflection factor $\underline{R}(\theta)$ shows a strong angular variation at the considered angular reflection angle θ_0. In the cases of ideally hard or soft boundaries (with $\underline{Z}_S \rightarrow \infty$ or $\underline{Z}_S = 0$ respectively) the model even yields an exact solution to the wave problem, since in these special cases the reflection factor is independent of the angle of incidence and thus the mirror source omnidirectional.

Figure 3.3 shows a quantitative assessment of this error for a locally reacting boundary, by plotting the percentage error $E = {}^{|\underline{p}^*_{\text{refl}} - \underline{p}_{\text{refl}}|}/_{|\underline{p}_{\text{refl}}|} \cdot 100$ between equations 3.12 and 3.13 as a function of the sum of heights of Q_0 and P and the angular reflection angle θ_0 for three different exemplary surface impedances $\underline{Z}_S$. The impedance values are taken from the paper by Suh and Nelson [1999] and span a wide range of different absorption levels.

It should be mentioned at this point that a similar quantitative assessment of the error of the mirror source approximation is given in the paper by Suh and Nelson [1999], who plot the error as a continuous function of θ_0 for the same three surface impedances $\underline{Z}_S$ and for three different normalized radial distances $\frac{r'}{\lambda}$. However, it is important to note the differences between our result presentation and the one by Suh and Nelson with regard to the independent variables used for the error determination. While Suh and Nelson chose $(\theta_0, r', \underline{Z}_S)$, we choose $(\theta_0, z' + h_s, \underline{Z}_S)$. With regard to room acoustic considerations we believe that it is more instructive to use the sum of heights $(z' + h_s)$ than the radial distance r'. This can for example be explained by considering that the plots by Suh and

Nelson misleadingly imply that the error is always small close to normal sound incidence but this result is based on the fact that for a given constant value of r' the sum of heights gets very large near normal incidence. And it is actually the large sum of heights that makes the error small rather than the angle of incidence as can be seen from figure 3.3.

In summary, the results in the figure 3.3 corroborate the findings by Mechel [2002] and it can be concluded that the image source approximation yields an acceptable error irrespective of the surface impedance and angle of incidence if the sum of heights above the reflecting boundary is greater than a few wavelengths. It is further interesting to note, that the overall error rises with increasing absorption. However, it appears that a general tendency for the angle dependence of the error cannot be given independent of the considered surface impedance at the boundary.

The modeling of the reflection of spherical sound waves at a planar boundary by a mirror source[3] is the basis of geometrical room acoustics simulations. This topic will be addressed in more detail in section 3.3.3, where a link between wave and geometrical room acoustics is established and also in section 4.2.1 where the image source method (ISM) will be described in the context of room acoustic simulations.

[3]The term image source is equally applied in the literature and both terms will be used equivalently throughout this thesis.

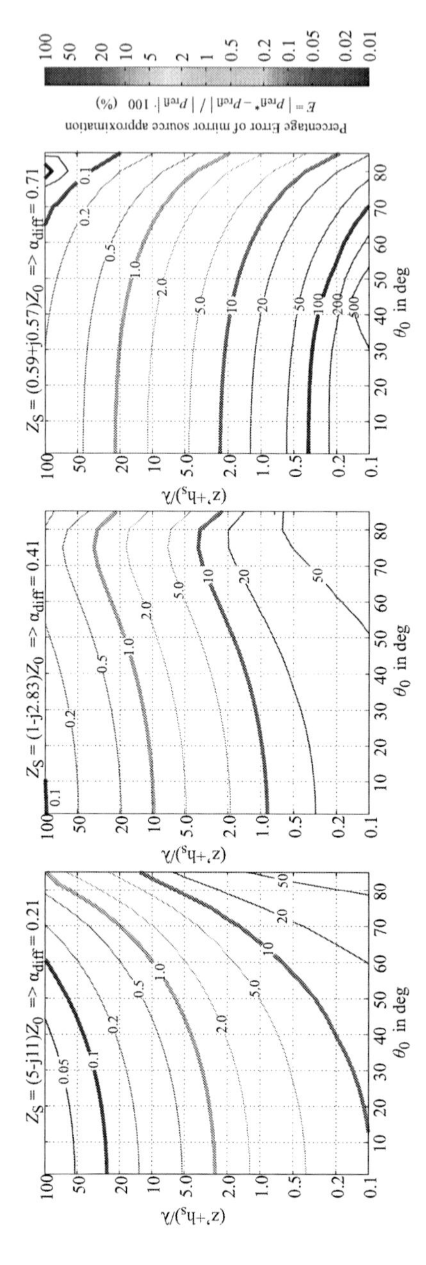

Figure 3.3.: Percentage error of the mirror source approximation (eq.: 3.13) compared to the exact solution (eq.: 3.12). The error is plotted as a function of the sum of heights $(z' + h_s)$ and the specular reflection angle θ_0 for three different impedances, which correspond to diffuse field absorption levels of 0.21, 0.41 and 0.71 respectively.

3.3. Room acoustics

3.3.1. Wave acoustics in rooms

Building up on the derivations given in section 3.1, the sound field in an enclosed space generated by an arbitrary sound source can thus be described as a solution to the inhomogeneous Helmholtz equation 3.6 with suitable boundary conditions on all room surfaces. From a mathematical point of view such boundary conditions for the Helmholtz equation are generally given by prescribing the sound pressure $\underline{p}$, its gradient $\nabla \underline{p}$ or a combination of the two at the boundary surfaces of the considered fluid domain. These conditions are known as Dirichlet, Neumann or mixed boundary conditions [Ehlotzky, 2007] respectively. In order to better understand how these mathematical boundary conditions relate to the acoustic conditions in real rooms, we want to consider the sound field in an arbitrarily shaped room with a loudspeaker source in it. In such a room the sound field is excited by the vibration of the loudspeaker membrane and the emitted sound waves which are traveling through the room get absorbed over time at the room boundaries and partly also by air-absorption, which shall be neglected for the moment. Since it was shown earlier that in the case of a locally reacting boundary surface its sound reflection characteristics can be fully modeled by the acoustic surface impedance $\underline{Z}_{\mathrm{S}}$, it can be concluded that the particular sound field in the room can be described by the Helmholtz equation with conditions for the velocity of the loudspeaker membrane and the acoustic surface impedances $\underline{Z}_{\mathrm{S}}$ at the room boundaries, which define the geometry of the room. Coming back to the mathematical conditions and taking into account that for time harmonic signals equation 3.1 (a) yields a proportional relation between the surface normal derivative $\frac{\partial \underline{p}}{\partial \boldsymbol{n}}$ and the surface normal velocity $\underline{v}_{\mathrm{n}}$ at a boundary, it is possible to show that prescribing the surface normal velocity on a boundary corresponds to a form of the Neumann condition as follows:

$$\frac{\partial \underline{p}}{\partial \boldsymbol{n}} = -j\omega\rho_0\underline{v}_{\mathrm{n}} \tag{3.14}$$

Similarly, a special form of a mixed boundary condition, which is also often referred to as 'impedance' or 'Robin' boundary condition, can be given in the following form:

$$\underline{Z}_{\mathrm{S}} = \frac{\underline{p}}{\underline{v}_{\mathrm{n}}} \qquad \Rightarrow \qquad \frac{\partial \underline{p}}{\partial \boldsymbol{n}} = \frac{-j\omega\rho_0}{\underline{Z}_{\mathrm{S}}} \cdot \underline{p} \tag{3.15}$$

A detailed discussion of these boundary conditions will be given in section 4.1 which deals with the finite element method in room acoustics. The following considerations will however be confined to a more general view on the nature of the solution of the Helmholtz equation in an enclosed space.

Although analytical solutions to the Helmholtz equation with given boundary conditions can only be given for a small number of idealized room shapes and boundary conditions, it can be shown [Kuttruff, 2000] that irrespective of the room geometry and boundary conditions these equations can generally only be solved for particular complex-valued, discrete eigenfrequencies $\underline{s}_m = \omega_m + j\delta_m$. These eigenfrequencies are each associated with a respective eigenmode function $\underline{p}_m(\boldsymbol{x})$ which corresponds to a characteristic spatial pressure distribution. Although the actual values of the eigenfrequencies and eigenmode functions strongly vary depending on room size, shape and boundary conditions, it can be stated that a sound field in a room is generally constituted by the superposition of the contributions from all these eigenmodes.

In the particular case where the room is excited by a simple point source at position $\boldsymbol{x}_0$, Kuttruff [2000] has shown that the frequency transfer function $\underline{p}(\omega, \boldsymbol{x})$ to a position $\boldsymbol{x}$ inside the room can in good approximation always be given in the following form:

$$\underline{p}(\omega, \boldsymbol{x}) = \sum_m \frac{\underline{A}_m}{\omega^2 - \omega_m^2 - j2\delta_m\omega_m} \tag{3.16}$$

where the coefficients $\underline{A}_m$ depend on the source position, the receiver position, the source strength and the frequency f. Under the reasonable assumption that the damping constants δ_m are small compared to the corresponding frequencies f_m, the absolute value of each series term can be simplified to

$$|\underline{p}_m(\omega, \boldsymbol{x})| = \frac{|\underline{A}_m|}{\sqrt{(\omega^2 - \omega_m^2)^2 + 4\omega^2\delta_m^2}} \tag{3.17}$$

which corresponds to the well-known frequency transfer function of a damped mass spring resonator system. In time domain such a resonator is represented by an exponentially decaying sinusoid of frequency f_m and decay constant δ_m. Thus, according to this formula a room can be considered as a kind of multiple resonator system, whose frequency transfer function can be described as the superposition of a large number of room eigenmodes at different frequencies f_m, which are in the context of room acoustics often referred to as 'standing waves'. This descriptive term stems from the fact, that each single eigenmode has its own spatially invariant pressure distribution (mode shape) $\underline{p}_m(\omega, \boldsymbol{x})$ that oscillates in time with the eigenfrequency f_m and decays exponentially with a damping constant δ_m. While the eigenfrequencies f_m and mode shapes $p_m(\boldsymbol{x})$ mainly depend on the overall room size and shape, the damping constants δ_m relate in a complicated way to the room size and the acoustic absorption characteristics at the room boundaries [Bistafa and Morrissey, 2003].

It is now interesting to investigate how the eigenfrequencies f_m are distributed along the frequency axis. For a rectangular room it can be easily shown that the average spacing between adjacent eigenfrequencies is given as a function of the frequency f as

$$\langle\frac{\mathrm{d}f}{\mathrm{d}N_\mathrm{f}}\rangle = \frac{c^3}{4\pi V f^2} \tag{3.18}$$

where the notation $\langle\ldots\rangle$ is used to denote the average of a quantity. It can be shown [Kuttruff, 2000, p.70] that for high frequencies this relation also holds in good approximation for non rectangular rooms. Thus the density of eigenfrequencies increases with f^2. Taking further into account that the average half power bandwidth of each eigenmode is proportional to the modes damping factor

$$\langle(2\Delta f)_n\rangle = \frac{\langle\delta_n\rangle}{\pi}, \tag{3.19}$$

it becomes clear that above a certain frequency the bell shaped resonance curves (cf. equation 3.17) corresponding to the different eigenmodes will start to heavily overlap on the frequency axis. In this frequency range the pressure amplitude at a given frequency f is thus given as the sum of contributions from many mutually independent eigenmodes and it can be shown by application of the central limit theorem of probability theory that the pressure amplitude at a given frequency f follows a Rayleigh distribution [Kuttruff, 2000]. On the other hand, at low frequencies where the distance between adjacent eigenmodes is in a similar range or even larger than their respective half power bandwidth, the pressure amplitude is dominated by just a few isolated eigenmodes. Further statistic properties of a room transfer function in the range of high modal overlap have been investigated by Schroeder [1954a]. His mathematical derivations regarding the average spacing and height of maxima in a frequency response, the average amplitude fluctuation and the average phase progression could also be verified experimentally [Kuttruff and Thiele, 1954].

The upper part of figure 3.4 illustrates how the impulse response and frequency transfer function can be considered as a multiple resonator system with increasing modal density on the frequency axis. Furthermore it shows how the pressure amplitude at a certain frequency with high modal overlap is calculated from several modal contributions. The transition between the modally and statistically dominated part of a room transfer function is given at the frequency where the spacing between the modes becomes so close, that several eigenfrequencies fall within the average half power bandwidth of room modes in this frequency range. In his original publication Schroeder [1954a] postulated that the statistical assumptions leading to the Rayleigh distribution of the pressure amplitude are fulfilled in good approximation if ten eigenmodes fall within a modes' average half power bandwidth. Based on extensive measurement studies Schroeder and Kuttruff [1962] later lowered this requirement to only three-fold modal overlap which results in the formula for the famous 'Schroeder frequency':

$$f_S = 2000\sqrt{\frac{T}{V}}, \quad \text{with} \quad T = \frac{6.9}{\langle \delta_n \rangle}, \tag{3.20}$$

where the influence of the damping factors on the average half power bandwidth is accounted for by means of the more commonly used room acoustic parameter 'reverberation time' T, which will be introduced in section 3.4.1.

3.3.2. Ray acoustics in rooms

In the limiting case where the overall dimensions of a room and its reflecting surfaces are large compared to the considered wavelength of a sound wave, it is admissible to model sound waves with the concepts of ray or geometrical acoustics [Kuttruff, 2000]. In this model a sound ray can be considered as a small portion of a spherical wave with infinitesimal aperture which originates from a certain source point and travels on a straight line at the speed of sound c. An impulsive sound source can thus be modeled as a point from which many sound rays are emitted at time $t = 0$. Furthermore the reflection of sound rays at room boundaries can be modeled by suitable specular or diffuse reflection laws, which account for the fact that a part of the ray's energy is absorbed at the wall reflection and the other part of the energy is reflected back into the room. Using these source and reflection models a room impulse response can thus be modeled by finding the valid reflection paths on which sound rays that originate from the source at $t = 0$

hit a designated receiver. Each of these reflection paths corresponds to a delayed impulse where its arrival time mostly depends on the path length and its amplitude and shape depend on the path length, the absorption characteristics at the room boundaries and the energy losses in the fluid medium. The sum of all reflections arriving after the direct sound is generally referred to as the reverberation of the room.

It is now interesting to make some general considerations regarding the temporal distribution of the reflection density and the decay characteristics of an impulse response, which is generated based on the concepts of geometrical acoustics. Under the simplifying assumption of a rectangular room with only specular reflections at the room boundaries, it can easily be shown that the average spacing between subsequent reflections arriving at a receiver is given by [Cremer and Müller, 1982]

$$\Delta t = \frac{V}{4\pi c^3 t^2}. \tag{3.21}$$

Later Vorländer [1995] has shown that this relation also holds in good approximation for non rectangular rooms. However, while the density of reflections increases with t^2 the intensity of a bundle of sound rays decays by $\frac{1}{t^2}$ since the sound pressure in an outward going spherical wave follows the well-known $\frac{1}{r}$ law (cf. equation 3.7). Thus, under the assumption that the contributions from overlapping reflections in a time interval $t + \Delta t$ can be considered incoherent, it is admissible to add up the intensities from all contributions [Kuttruff, 2000], which leads to a mutual compensation of the decrease in intensity and the increase in reflection density in a time interval $t+\Delta t$. Consequently, the temporal energy decay in a room impulse response is only caused by the energy loss at the wall reflections and the propagation losses in the fluid medium. By further relating the average number of wall reflections after time t to the mean free path $\bar{d}$ in the room [Cremer and Müller, 1982] the following energy decay law can thus be derived:

$$E(t) = E_0 \exp\left(-mct + \frac{ct}{\bar{d}} \ln(1 - \alpha)\right), \quad \text{with} \quad \bar{d} = \frac{4V}{S}, \tag{3.22}$$

where E_0 is the initial energy at $t = 0$, $\frac{ct}{\bar{d}}$ is the average number of wall reflections after time t, m is the absorption constant of the medium air, V is the room volume and S the room surface respectively. Furthermore $\bar{\alpha}$ is the average absorption coefficient of the walls. It is important to note that equation 3.22 only holds if the mean free path is a good approximation of the average distance between two reflections for all sound rays traveling through the room. In other words the sound field in the room must be sufficiently diffuse. Deviations from this idealized decay curve are often perceived in an unpleasant way and can for example be caused by flutter echoes or sound focusing from concave surfaces.

3.3.3. Links between wave and ray theory

At first sight it appears difficult to establish a link between the representations of a room acoustic transmission path based on wave and geometrical acoustics. While in wave acoustics the room impulse response is modeled as the superposition of many exponentially decaying harmonic eigenmodes, geometrical acoustics models the room impulse response as a temporal succession of sound pulses reaching a receiver over many different reflection paths. These different views are illustrated in figure 3.4. However, in their landmark paper of 1979 Allen and Berkley were able to show that for a rectangular room with rigid boundaries which is excited by an ideal point source the results from a modal superposition based on the solution of the Helmholtz equation and those from a geometrical mirror source approach[4] as indicated in section 3.2.3 are indeed equivalent. This can be understood, by taking into account that the impulse response generated by a geometrical mirror source model consists of the superposition of simultaneously emitted impulses from the original point source and many mirror sources outside the room. Thus the resulting sound field that is generated inside the room must satisfy the wave equation, since the sound field of each point source does so. Taking further into account that the mirror source positions are constructed in a way such that the trivial boundary condition for a rigid surface $v_n = \frac{\mathrm{d}p}{\mathrm{d}n} = 0$ is satisfied on each boundary it becomes clear that the resulting sound field inside the room has to yield the unique solution for the given boundary problem of the Helmholtz equation. Sections 4.2.1 and 7.1, which deal with the image source method in detail, will discuss the implications of this important result in the case of arbitrarily shaped rooms with inhomogeneously distributed and frequency dependant absorbing boundaries. For the moment it shall only be emphasized that also in more complex rooms the temporal distribution of reflections in the time domain contains important information regarding the modal structure of a room transfer function and vice versa.

More interesting parallels between wave and ray acoustics can be found when looking at the distribution and shape of eigenmodes in frequency domain and reflections in time domain. It is for example striking that the derivations of the average modal spacing (cf. eq. 3.18) and the average spacing between successive reflections (cf. eq. 3.21) lead to analog expressions in time and frequency domain. Moreover, in both domains we find a transition from a deterministic part for low frequencies (early reflections) to a stochastic part with high modal (reflection) overlap and an increase of the damping in the room leads to a widening of the bell-shaped eigenmodes in frequency domain and of the mirrored reflections in time domain. Schroeder [1996] stated in this context that the transition frequency f_S and the transition time t_trans between the deterministic early reflection and the stochastic diffuse field regime describe the same physical phenomenon in two different spaces and are related by the simple constant factor:

$$t_\mathsf{trans} = \sqrt{\frac{3}{8\pi}\frac{1}{f_\mathrm{S}}}. \tag{3.23}$$

While the room acoustic literature is very consistent with regard to the usage of the Schroeder frequency as the transition frequency between the deterministic and stochastic modal regime, the transition in time domain is treated much less consistently in the literature and many other definitions of the transition time t_trans have been suggested. A nice overview of different measures for this temporal transition is given by Jeong et al. [2010].

[4] The mirror source approach was discussed for a single reflection at a planar boundary in section 3.2.3. In the case of a rectangular room an infinite number of image sources can be constructed by recursively mirroring each mirror source again on all walls except the parent wall. This will be discussed in detail in section 4.2.1

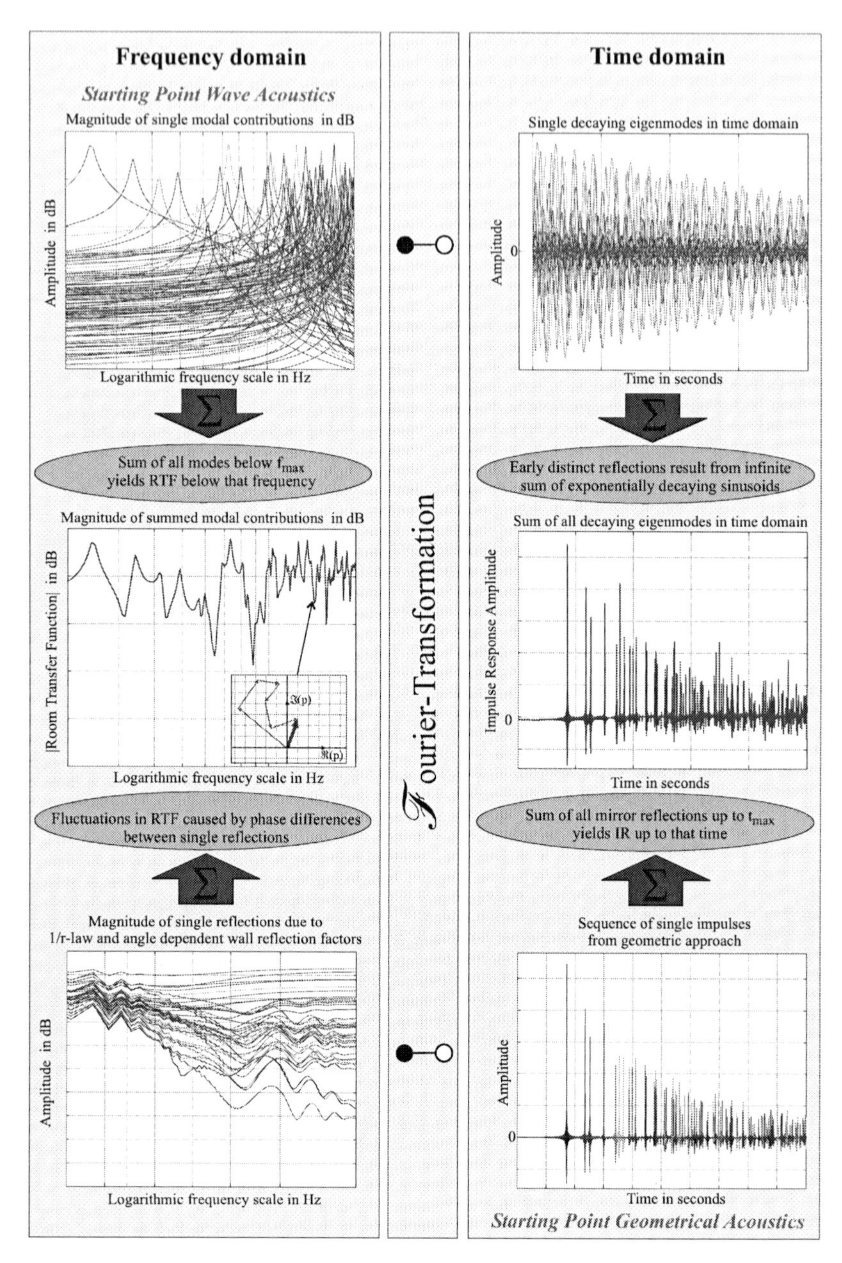

Figure 3.4.: Time and frequency domain representation of the sound transmission path in a room based on wave and geometrical acoustics considerations.

Summing up it can be concluded, that besides the analogies in time and frequency domain the relation between reflections in time domain and room modes in frequency domain remains complicated and difficult to grasp, since strictly speaking each room mode contributes to an infinite number of sound reflections and on the other hand each sound reflection contributes to an infinite number of room modes (cf. figure 3.4).

3.4. Objective and subjective evaluation of room acoustics

It has so far been explained that the sound transmission between a given source and receiver in a room can be equivalently described by the room transfer function in the frequency domain and the room impulse response in the time domain. Thus, any room acoustic simulation tool, which aims at the auralization of the sound field in an already existing or planned room, needs to replicate/predict the room transfer function/impulse response or at least its key perceptual features to the best possible extend. While for the modally dominated part of the sound field a direct comparison of the fine structure of the measured and simulated room transfer functions appears reasonable, such a direct comparison appears both hopeless and senseless at frequencies far above the Schroeder frequency, due to the statistical fluctuations induced by high modal and reflection overlap. In this context Schroeder [1954b] has also shown that in the frequency range with high modal overlap very small deviations from the original room setup lead to completely random changes in the statistics of the eigenmode frequencies and their excitation strength. Thus, with regard to a small room simulation the presence of a listener in the room, a small deviation from the original geometry or an error in the determination of the boundary conditions will already completely randomize the eigenmode statistics compared to the empty real room. For the statistical part of the RTF/RIR room acousticians therefore generally resort to integral, averaged measures which aim at capturing the key perceptual characteristics of the room sound field.

The present section therefore introduces these objective measures with a focus on the reverberation time (T) and the energy decay curves (EDC). Moreover, we will touch on some important issues, related to the application of classical room acoustic parameters in small room acoustics and finally discuss the difficulties in assessing the quality of room acoustic simulations without subjective listening tests.

3.4.1. Temporal sound decay and reverberation time

Already in 1923 Sabine stated that the average decay rate of a sound in a room is one of the most important characteristics in the evaluation of the acoustical quality of a room and thus defined the reverberation time T of a room as the time in which the sound pressure level of an interrupted sound decays by an amount of 60 dB below its steady state level. The value of 60 dB is motivated by the fact that for moderate excitation levels the room reverberation becomes inaudible at this attenuation level. Based on his experimental findings he derived a simple prediction formula which related the reverberation time to the volume V and total absorption A of a room. In 1930 Eyring derived a refined formula for the reverberation law, which gives more accurate predictions in rooms with considerable absorption. The Sabine (T_S) and Eyring (T_E) reverberation laws are given as follows:

$$T_\mathrm{S} = 0.161\frac{V}{A}, \quad \text{with} \quad A = mc + S\bar{\alpha} \tag{3.24}$$

$$T_\mathrm{E} = 0.161\frac{V}{mc - S\ln(1-\bar{\alpha})}, \tag{3.25}$$

where S is the room surface, m is the absorption constant of the medium air and $\bar{\alpha}$ is the average absorption of all room boundaries. Although not explicitly stated by Sabine, the influence of the absorption constant of air m in the formula for T_S is included here for the sake of completeness. For low values of $\bar{\alpha}$ both prediction formulas yield almost identical results.

A method to determine the reverberation time of a room based on its room impulse response is described in ISO 3382-1:2009. In this method the interrupted steady state sound decay is mimicked by a backward integration of the squared room impulse response as follows:

$$D(t) = 10\log_{10}\left(\int_t^\infty h^2(\tau)\,\mathrm{d}\tau\right), \tag{3.26}$$

where $h(t)$ is the impulse response and $D(t)$ is the level of the backward integrated sound decay. Under the assumptions of an almost linear decay of $D(t)$ over time the reverberation time is calculated from the slope coefficient β of a linear regression of the decay curve on the time t as follows:

$$T = -\frac{1}{\beta}\cdot 60\,\mathrm{dB}. \tag{3.27}$$

Since in practical measurements it is not always possible to use the full 60 dB decay for the linear regression (due to an insufficient signal-to-noise ratio), the slope coefficient is often calculated using only a restricted part of the sound decay. The according reverberation times, which are most commonly used in practice, are thus denoted T_{20} (if the slope coefficient is determined based on the decay between -5 and -25 dB) or T_{30} (-5 and -35 dB).

The frequency dependence of the sound decay is generally captured by bandpass filtering the impulse response using octave band or third octave band filters and then carrying out the backward integration and linear regression for all single frequency bands. It is however important to note, that both the applicability of the Sabine and Eyring prediction formulas as well as the assumption of a linear sound decay rely on sufficient diffusity of the sound field in the considered room. It can be shown that especially at low frequencies deviations from this idealized, linear sound decay are often caused by contributions from strongly excited room eigenmodes with decay times that considerably exceed the average decay times. Due to an insufficient number of eigenmodes in the frequency bands below f_S the contributions from these modes might not average out. Additionally, in rooms with strong flutter echoes, focusing effects or very inhomogeneously distributed absorption these deviations from the idealized, linear sound decay may also persist in the higher frequency ranges. In such cases it is therefore more instructive to directly compare the band-filtered energy decay curves (EDC) instead of only looking at the reverberation times.

3.4.2. Evaluation methods used in this thesis

Besides the reverberation time T, the ISO 3382-1:2009 standard defines more objective room acoustic parameters such as the 'Clarity' C, 'Definition' D or 'Sound Strength' G, which can also be derived on the basis of a monaural room acoustic impulse response. Other important parameters are for example related to speech intelligibility like the 'Speech Transmission Index' (STI) and its modified version the 'Rapid Speech Transmission Index' ($RASTI$) (cf. DIN EN 60268-16). While all these parameters offer valuable insight into the acoustic characteristics of moderately or highly reverberant large spaces, they often appear hardly insightful in the case of acoustically interesting small rooms, like recording studios or car passenger compartments, with generally very short reverberation times and a much higher reflection density. For example, in the case of a car passenger compartment with an average mid frequency reverberation time of less than 0.1 s it is hardly reasonable to discuss room acoustic quality on the basis of these classical room acoustic parameters.

While it is not the aim of the present thesis to derive modified or even new objective measures for the evaluation and comparison of sound fields in small rooms, we inevitably have to face the problem that a comparison of simulated and measured room impulse responses based on a comparison of the existing objective room acoustic parameters, will supposedly not describe a large fraction of all relevant perceptual differences in the small room impulse responses. Moreover, it is doubted that existing difference limens for these parameters, which were generally obtained for larger rooms, can be applied straightforward to the range of values obtained in typical small rooms. The present thesis therefore does not report any of the aforementioned room acoustic parameters except for the reverberation time, where suitable. Instead, the objective evaluation of the modally dominated part of the simulated room transfer functions is based on a direct comparison with the corresponding measured low frequency RTFs. Comparisons of the stochastic part of the transfer function are based on band averaged energy spectra, since it was found that in rather 'dry' small rooms the RTF acts somehow like an equalizer which emphasizes some frequency bands and attenuates others. Additionally, in some cases, the reverberation time characteristics and the spectrograms of the considered room impulse responses are plotted and compared in order to also evaluate the time characteristics of the room impulse response.

Since it is very difficult to relate differences in the averaged or non-averaged transfer functions to perceptual differences in the impulse response, frequency and time dependent loudness curves are also reported for representative audio signals which were convolved with the measured and simulated impulse responses in chapter 8. The differences in the loudness curves are expected to be more closely related to human perception than those in the raw spectra and impulse responses. This notion is also supported by the work of Lee and Cabrera [2009], Lee et al. [2011] who introduce a room acoustic sound field evaluation based on the so-called loudness decay function. The authors also provide some evidence that this subjectively motivated measure is indeed more closely related to the sound experienced by the listeners than typically used objective, impulse response based measures. Finally, for the rooms in section 8 and 9 we also report preliminary results from subjective comparisons of auralizations based on measured and simulated impulse response data. A more thorough subjective evaluation and comparison of measured and simulated impulse responses is however strongly advised for future work and first steps into this direction have already been conducted by Aretz and Jauer [2010], who present a clear-cut terminology for the subjective benchmark of the quality of room acoustic simulations together with suitable listening tests. The suggested terminology and the corresponding listening tests distinguish three quality levels based on different requirements that may be imposed on room acoustic simulation results. A closer look at this terminology will be given in the outlook section 10.2 of this thesis.

4. Applied room acoustic simulation methods

Based on the theoretical fundamentals of room acoustics lain down in the previous chapter, the present chapter discusses the principles of the room acoustic simulation tools that have been applied in the course of this thesis. The chapter is organized as follows: Section 4.1 gives a condensed derivation of the finite element formulation in room acoustics. Next, section 4.2 discusses how the concepts of geometrical acoustics theory can be integrated into state-of-the-art simulation algorithms based on the image source (IS) and stochastic ray tracing (SRT) method, where special emphasis is put on the potential of the image source method to predict the modal low frequency structure of the sound field under certain conditions. Finally, section 4.3 deals with the combination of wave and ray based simulation results in the frequency domain to a single room acoustic impulse response.

4.1. Room acoustic FE simulations

The Finite Element Method (FEM) is a powerful tool for the numerical solution of partial differential equations with given boundary conditions and is widely used in numerous fields of engineering sciences. Typical applications in acoustics deal with the prediction of the modal characteristics of structure borne, airborne and also coupled sound fields in enclosed spaces. The aim of the present thesis is to use the FEM for low frequency room acoustic simulations in arbitrarily shaped rooms with various absorber configurations on the room boundaries. The following sections therefore give a very short introduction to the terminology and the fundamental equations of an FE formulation for room acoustics with all necessary source and boundary conditions. Finally, a list of requirements for the simulation model will be given that have to be met in order to obtain reasonable simulation results.

All FE simulations in the course of this thesis were carried out using either the proprietary FE solver *WAVE*, which has been developed at ITA of RWTH Aachen since 2002, and/or the commercial FE software *Virtual Lab*[1], and very good agreement was found between the results from both softwares. Since the solution of the FE system of equations is deterministic, this result is admittedly not surprising. However, by running simulations with both softwares for identical simulation setups, meshes and boundary conditions it was possible to minimize the risk of false simulation results caused by user errors. Moreover, by comparison with the *Virtual Lab* results it was possible to validate the proprietary solver *WAVE* for all different configurations of boundary and source conditions.

[1] http://www.lmsgermany.com/simulation/acoustics/finite-element-acoustics

4.1.1. Finite element fluid model

The wave propagation in a fluid medium in an enclosed cavity Ω can generally be described by the Helmholtz equation (eq. 3.4) with given impedance (Robin) boundary conditions (eq. 3.15) on the boundaries Γ_Z and velocity (Neumann) conditions (eq. 3.14) on the boundaries Γ_v of the fluid domain. Using the fundamental theorem of the calculus of variations [Knothe and Wessels, 1999, p.438ff] the governing equations of the fluid FEM can be given by writing the Helmholtz equation and the given boundary conditions in an equivalent integral form and multiplying by an arbitrary weighting function $\bar{w}$:

$$\int_\Omega \bar{w}\left(\Delta p + k^2 p + j\omega\rho_0 q\right) d\Omega + \oint_{\Gamma_v} \bar{w}\left(-\frac{\partial p}{\partial \boldsymbol{n}} - j\omega\rho_0 v_n\right) d\Gamma_v + \dots$$
$$\dots \oint_{\Gamma_Z} \bar{w}\left(-\frac{\partial p}{\partial \boldsymbol{n}} - \frac{j\omega\rho_0}{\underline{Z}_S} p\right) d\Gamma_Z = 0, \qquad (4.1)$$

where p is the sound pressure, ω the harmonic frequency, j the complex number, k the propagation constant, ρ_0 the fluid density and q the source term according to equation 3.6. $\underline{Z}_S$ gives the acoustic surface impedance (for normal incidence) and v_n the normal velocity at the corresponding fluid boundaries Γ_Z and Γ_v respectively. The essence of equation 4.1 can be summarized by simply noticing that if a function p observes the inhomogeneous Helmholtz equation and the boundary conditions on Γ_Z and Γ_v the integrands will be zero for any arbitrary weighting function $\bar{w}$ and thus equation 4.1 is fulfilled. Showing that equation 4.1 is indeed equivalent to the statement of the inhomogeneous Helmholtz equation with given boundary conditions (under certain conditions for the weighting functions $\bar{w}$) is the core of the fundamental theorem of the calculus of variations. Since showing the mathematical equivalence is a bit more difficult, interested readers are referred to standard books on the topic (e.g. Funk [1970]).

Based on the above integral equation the FE system of equations can then be derived in the following three steps:

1. By making use of Green's first identity[2] [Levandosky et al., 2008] it is possible to reduce the highest derivation order in equation 4.1 to order 1 and to eliminate the normal pressure derivatives in the boundary integrals, which yields:

$$\int_\Omega -\nabla\bar{w}\nabla p + \bar{w}\left(k^2 p + j\omega\rho_0 q\right) \mathrm{d}\Omega - \oint_{\Gamma_v} \bar{w}\left(j\omega\rho_0 v_n\right) \mathrm{d}\Gamma_v - \dots$$
$$\dots \oint_{\Gamma_Z} \bar{w}\left(\frac{j\omega\rho_0}{\underline{Z}_S} p\right) \mathrm{d}\Gamma_Z = 0, \qquad (4.3)$$

   It can be shown that using this equation considerably relieves the continuity constraints for the elementwise approximations to the pressure and weighting functions introduced in step 3.

[2]Green's first identity is given by:

$$\int_\Omega (\psi\Delta\varphi + \nabla\psi\nabla\varphi)\, \mathrm{d}\Omega = \oint_\Gamma \psi\left(\frac{\partial\varphi}{\partial\boldsymbol{n}}\right) \mathrm{d}\Gamma, \qquad (4.2)$$

where ψ and φ are scalar functions of which φ is twice continuously differentiable, Ω is a region in $\mathbb{R}^3$ with boundary Γ and $\boldsymbol{n}$ is the outward pointing surface normal on Γ.

2. In a second step the integration domain (volume and boundary surfaces) is subdivided into finite elements. This processing step is often referred to as "meshing" and transforms equation 4.1 into a sum of integrals over all finite elements.

3. In order to find an approximation to the pressure function that observes equation 4.3 it is further necessary to limit the solution space of the pressure and weighting functions in each element. In classical FEM this is generally done by using element-wise polynomial approximations (also called trial or test functions). In the course of this thesis we exclusively use 2nd order polynomials of the serendipity type [Zienkiewicz, 1977] for tetrahedral and hexahedral 3D fluid elements and triangular and quadrangular 2D elements for the boundary surfaces. In order to also allow curved element shapes, all elements are of isoparametric type [Zienkiewicz, 1977], which means that the geometry of an element is described by polynomial functions of the same order as the pressure and weighting functions. For the sake of completeness it should be mentioned that various other sets of functions exist. A very interesting recently developed technique called Wave Based Method (WBM) uses globally defined wave functions instead of the polynomial approximation thus reducing the required fineness of the mesh discretization. The Wave Based Method belongs to the family of so-called Trefftz methods and has been presented and successfully applied by Desmet and his co-workers in various publications [Desmet, 1998, Van Hal et al., 2003, Van Genechten et al., 2009].

   Irrespective of the chosen set of functions this means that for each element the continuous pressure function is fully described by the sound pressure at a finite number of nodal points within the element. By additionally enforcing continuity on the element boundaries, the search for a continuous pressure function is thus reduced to the search of a finite number of degrees of freedom describing the polynomial pressure function in each element. Moreover, the use of the polynomial trial functions for each element allows to write the elementwise integrals of equation 4.3 as a weighted sum of contributions from the discrete nodal mesh points, which makes it possible to write the integral FE equations in a closed form matrix notation.

Following steps 1 to 3 the following system of FE equations can be derived for the fluid domain:

$$\left(\boldsymbol{K}_F + j\omega\underline{\boldsymbol{A}}_F - \omega^2\boldsymbol{M}_F\right)\underline{\boldsymbol{p}} = j\omega\underline{\boldsymbol{f}}, \tag{4.4}$$

where the matrices $\boldsymbol{K}_F$ and $\boldsymbol{M}_F$ are called compressibility and mass matrix respectively, the damping matrix $\boldsymbol{A}_F$ comprises the boundary terms for the surfaces with the impedance conditions $\underline{Z}_{\mathrm{S}}(\omega)$ and the right-hand side vector $\boldsymbol{f}$ comprises the excitation of the sound field by the surfaces with given velocity $\underline{v}_{\mathrm{n}}(\omega)$ and the continuous volume velocity density function $\underline{q}(\omega)$. The pressure field in the fluid domain can then be calculated by inversion of the given system of equations. Extensive introductions to the FEM can be found in the renowned books by Zienkiewicz [1977] or Bathe and Zimmermann [2001]. However, these books generally derive the principles of the FEM based on the equations of structural dynamics, which are sometimes a little hard to digest for fluid acoustics people. For a detailed introduction to the FEM in fluid acoustics the dissertation by Franck [2008] is therefore recommended.

It can be shown that for an increasing fineness of discretization or alternatively an increasing order of the polynomial approximation the FE solution converges to the exact (analytic) solution for a given fluid domain and boundary conditions. However, in typical situations where we want to simulate the sound field in an already existing or planned room, we frequently face the problem that we have to identify realistic models for the boundaries of the fluid domain and that appropriate input parameters of these models have to be determined by measurement. The challenges of the determination of realistic boundary conditions will be dealt with at length in chapter 5, 8 and 9. The following sections give a summary of the required boundary input data and some general remarks on all factors that influence the quality of a room acoustic FE simulation including recommendations regarding the required level of detail in the CAD model, the mesh fineness and the frequency resolution.

4.1.2. Boundary, source and binaural receiver model in FE domain

The present section describes in detail how the boundary, source and receiver characteristics can be realistically modeled in an FE simulation. Furthermore it summarizes from a practitioners point of view the necessary input data that needs to be acquired to run an FE simulation. The later sections 5 and 6 will then deal with the measurement and calculation methods that can be used to acquire this data.

1. **Boundary representation**

   According to section 3.2.1 the reflection characteristics of a specific boundary surface in a room can be accounted for by the acoustic surface impedance $\underline{Z}_{\mathrm{S}}(\omega)$ of this boundary. As was already discussed earlier this model is strictly speaking only adequate in the case of locally reacting boundaries. It is therefore emphasized again that in a room acoustic FE simulation the consideration of a laterally reacting boundary with $\underline{Z}_{\mathrm{S}} = f(\theta)$, where θ is the angle of incidence on the boundary surface, is not possible by using an impedance boundary condition. This would necessitate an appropriate model of the sound propagation in the boundary domain with a mutual coupling of the fluid and boundary domain.

2. **Source representation**

   The consideration of loudspeakers in room acoustic FE simulations can be effectuated in two different ways, which can both be deduced from equation 4.1. Firstly a sound source can be modeled by assigning a surface normal velocity v_{n} to a boundary surface Γ_{v}. This would for example be adequate for a loudspeaker which is mounted flush into a room wall where Γ_{v} is the effective membrane surface and v_{n} the surface normal membrane velocity. If the loudspeaker is standing inside the room, this source representation would of course also require the modeling of the loudspeaker cabinet as a boundary in the FE domain. Since the directional characteristics of a loudspeaker source are mostly caused by diffraction at the loudspeaker cabinet and in case of a multichannel loudspeaker interference between the different driver membranes, such a model of the loudspeaker cabinet with driver units (membranes) would then be able to fully capture the directivity characteristics of this loudspeaker. However, since FE simulations are generally conducted for the lower frequency range, where most loudspeakers have an almost omnidirectional radiation pattern, it is sometimes reasonable to substitute the full geometrical loudspeaker model (cabinet with radiating membranes and bass reflex holes) by a simple point source model, which is fully described by its volume velocity Q (cf. eq. 3.7). In the FE formulation such a point source can be considered by setting the volume velocity density function q in equation 4.1 to a spatial dirac impulse with source strength Q. In the right hand side excitation vector $\underline{f}$ of the FE system of equations this results in excitation terms for all mesh nodes belonging to the volume element that comprise the point source.

3. **Binaural Receiver representation**

   Since the solution of the FE system of equations yields the sound pressure at every position in the room (even at positions where no mesh node is present, since the pressure function between the mesh nodes is determined by the polynomial trial functions used for each element) no special receiver model is required in room acoustic FE simulations. On the other hand it is important to mention, that the consideration of a binaural receiver would strictly speaking necessitate a model of the listeners head and torso in the FE simulation. This is however only reasonable if either the room volume is so small that the inclusion of the head and torso considerably affects the modal structure of the sound field and/or if the considered frequencies are so high that interaural differences between the left and right ear can no longer be neglected in good approximation, which is roughly above 500 Hz [Fels, 2008]. This problem will be discussed in detail in section 9 when dealing with the FE simulations in a car passenger compartment.

4.1.3. Requirements on the room model, meshing and frequency resolution

Apart from the determination of suitable input data for the boundary and source conditions for the room acoustic FE simulation further requirements related to the geometrical room model and mesh as well as to the discrete frequency resolution of the simulation need to be considered which are summarized in the following:

1. The FE mesh needs to represent all relevant details in the geometry of the simulation domain well. This means that the model has to incorporate every detail that is not small compared to the shortest wavelength considered in the simulation. This has obviously some implications on the fineness of the required mesh and thus the computation time.

2. The mesh discretization must be sufficiently fine and/or the polynomial order of the element functions sufficiently high to realistically represent the wave character of the sound field. As a rule of thumb a minimum of three elements per wavelength is required to realistically represent the wave character of the sound field for a second order polynomial approximation in each element [Thompson and Pinsky, 1994]

as used throughout the present thesis. For these elements the following meshing constraint can thus be derived as a function of the highest simulation frequency:

$$l_{\text{Element}} \leq \frac{\lambda_{\min}}{3} = \frac{c}{3f_{\max}} \tag{4.5}$$

As a consequence the pressure degrees of freedom of the model increase proportional to $f_{\max}^3$.

3. The system of equations is solved for discrete frequency steps, and hence the frequency resolution Δf has to be specified by the user. Since the total computation time is directly proportional to the number of frequency steps a trade-off has to be made in order to obtain a sufficient density in the frequency domain whilst at the same time avoiding excessive computation times. A first constraint for the Δf can be derived on the fact that in discrete Fourier analysis the frequency resolution is directly related to the length of the signal in the time domain by $\tau_{\text{RIR}} = \frac{(N-1)}{f_{\text{sampling}}}$, where τ_{RIR} is the length of the room impulse response (RIR), N is the number of samples and f_{sampling} is the sampling frequency. Assuming that the length of the RIR should be at minimum equal to the expected reverberation time T in the simulated room and using $\Delta f = \frac{f_{\text{sampling}}}{N}$, the first criterion for the frequency resolution is given by:

$$\Delta f \leq \frac{1}{T} \tag{4.6}$$

A second criterion can be derived by considering the simulation phase. The frequency steps have to be chosen sufficiently close in order to assure that the phase can be unwrapped correctly. This is made sure by requiring that the phase forward $|\Delta\varphi|$ between two subsequent frequencies should not exceed 180° or π respectively. Theoretical formulas for the average phase forward per Hz are given by Schroeder [1959] for the extreme cases of no modal overlap and high modal overlap:

$$-\left\langle \frac{d\varphi}{df} \right\rangle_{\text{modal}} \approx \frac{4\pi^2 V f^2}{c^3}, \quad \text{no modal overlap} \tag{4.7}$$

$$-\left\langle \frac{d\varphi}{df} \right\rangle_{\text{statistic}} \approx \frac{T}{2.2}, \quad \text{high modal overlap} \tag{4.8}$$

A sufficient condition for the frequency resolution Δf can now be derived by satisfying $|\Delta\varphi| < \pi$ at the steepest position of $|\frac{\mathrm{d}\varphi}{\mathrm{d}f}|$ according to equations 4.7 and 4.8. Taking equation 4.7 as an approximation of $|\frac{\mathrm{d}\varphi}{\mathrm{d}f}|$ below the Schroeder frequency f_{S}, and equation 4.8 for frequencies above, it can be shown that the steepest position of $|\frac{\mathrm{d}\varphi}{\mathrm{d}f}|$ is given by equation 4.7 at the Schroeder frequency. A sufficient condition for Δf can therefore be derived by writing $|\frac{\mathrm{d}\varphi}{\mathrm{d}f}|$ from equation 4.7 as a differential quotient $\frac{\Delta\varphi}{\Delta f}$ and imposing that $|\Delta\varphi| < \pi$ for $f = f_{\text{S}}$. This second criterion, which is more restrictive than 4.6, can be given as follows:

$$|\Delta f| < \frac{c^3}{4\pi V f_{\text{S}}^2} = \frac{0.8}{T} \tag{4.9}$$

4.2. Hybrid geometrical acoustics simulations

Nowadays, most state of the art geometrical acoustics simulation tools use hybrid algorithms that combine an image source (IS) method for a precise calculation of the early specular reflections in a room impulse response (RIR) with a computation efficient stochastic ray tracing (SRT) algorithm to calculate the late diffuse exponential sound decay. Whilst implementation details of the IS and SRT algorithm might vary considerably among different simulation tools, most have in common that the fine structure of the IR is split up in an early deterministic part based on the ISs and a late stochastic part calculated from the SRT results. The following sections give some general considerations regarding the IS and SRT methods, without going into the implementation details of a specific software. Special focus is put on the potential of the image source method in predicting the modal structure of the sound field.

All geometrical acoustics simulations presented in the course of this thesis were conducted using the simulation software *RAVEN*, which is currently developed at ITA of RWTH Aachen University. An extensive description of the working principle and features of the simulation software *RAVEN* as well as a validation of the implemented algorithms is given in the dissertation by Schröder [2011].

4.2.1. Image source method (ISM)

The ISM is based on the principle that the sound field generated by a point source in front of an extended planar surface can be represented in good approximation by the superposition of the sound field generated by the original point source and that generated by an additional secondary source, called image source. This image source is constructed by mirroring the original point source at the reflecting surface and weighting its amplitude by the complex reflection factor $\underline{R}(\theta)$ of the considered boundary surface (cf. section 3.2.3). In a room acoustic scenario where the flat room surfaces can be considered as fragments of extended planes such image sources can be constructed by mirroring the original source at all planes belonging to the boundary surfaces of the room. These sources are called image sources of first order. Higher order image sources, which represent sound paths over multiple wall reflections can then be constructed by mirroring the image sources again and again on all planes belonging to the room boundaries. It has to be noted that depending on the shape of the room and the position of the receiver not all of these image sources represent valid sound path and therefore a visibility check needs to be carried out for all image sources. A quite recent and comprehensive discussion of the image source method with regard to the geometrical construction of the sources and the required criteria for their visibility is given by Mechel [2002].

It was already mentioned earlier (cf. section 3.3.3) that in the case of a cuboid room with ideally rigid boundaries the image source method converges to the solution of the Helmholtz equation [Allen and Berkley, 1979]. Following the reasoning by Mechel [2002] this fundamental result can be extended to strictly concave rooms ($\theta_{corner} < 180°$) with either ideally soft ($\underline{Z}_S = 0$) or ideally hard ($\underline{Z}_S \to \infty$) boundary conditions. However, it can not be generalized to arbitrary room shapes and boundary conditions both for theoretical and practical reasons. The theoretical reasons are related to the following two main facts:

1. The image source model can, in its original form, not model the sound reflection at convex edges or corners. Although extensions to the image source method that account for diffraction at convex edges have been suggested by Svenson et al. [1999] or Mechel [2002] these extensions considerably increase the computational complexity of the method and are hardly applicable for complex shaped rooms with a very high number of surfaces and diffraction edges. Moreover the image source method cannot adequately model the sound reflection at a flat surface that consists of patches with very different reflection characteristics, since the diffraction effects at the joints of the different material patches are not considered.
2. In a strict sense the sound field generated by the mirrored point source with directivity $\underline{R}(\theta)$ violates the wave equation. The generated error is however considered negligible in the case of moderate angles of incidence $\theta_{inc} < 60°$ [Suh and Nelson, 1999] or as Mechel [2002] puts it in the case of "a great sum of heights of Q and P over the wall", where Q and P are the source and receiver positions respectively (cf. section 3.2.3).

On the other hand important practical reasons that limit the accuracy and applicability of the image source method in real rooms are mostly related to the computational challenge of finding all valid image sources up to high orders in room models with a large number of surfaces and to the problem that sound scattering at objects with high geometrical detail (such as furniture objects) cannot be adequately modeled in the ISM model. This is due to the fact that on the one hand the number of planar surfaces would explode to a non-manageable level, if all room details were approximated by planar sub areas, and that on the other hand these scattering objects would consist of a very high number of convex diffraction edges which cannot be handled by the ISM anyway.

In large room acoustics it is widely accepted that the scattered energy typically dominates the specularly reflected energy in an ordinary room already from order $n \geq 3$ [Kuttruff, 1995] and thus the late stochastic part of the IR is generally more adequately modeled using an SRT algorithm (see section 4.2.2), which generally uses a rough sound scattering approximation. However, this result by Kuttruff cannot be generalized to the modally dominated low frequency range in small rooms, where the sound field is dominated by a few fundamental eigenmodes which are mostly determined by the overall room shape and are only marginally affected by small geometric details in the room. It is thus appropriate to ask the question if the ISM can be used to predict the modal part of the sound field in real rooms with a basically concave shape using a simplified room model and a reasonable number of image source orders. This interesting question shall be investigated in detail in section 7.1.

4.2.2. Stochastic ray tracing (SRT)

SRT algorithms model the propagation of sound waves from a given source to a receiver by means of sound energy particles. These particles propagate on rays that are emitted from a point source into random directions whereby each particle carries a discrete proportion of the total energy of the source. A proportion of the particle's energy is dissipated each time a particle hits a reflecting surface. In contrast to the IS method the SRT method also accounts for sound scattering at a boundary where the amount of scattered energy is determined by the so-called scattering coefficient s which is defined as the ratio between the non-specularly and the total reflected energy emitted from a boundary surface. The angular distribution of the scattered energy is typically chosen according to Lambert's cosine law which provides an adequate model of the sound scattering at an ideally diffuse reflecting surface. The Lambert cosine law is adapted from the field of optics and is implemented in most acoustic ray tracing implementations. The sound receiver is represented by a detection sphere counting all crossing particles and sorting the impacting energy into a time dependent energy histogram, which can be interpreted as the energy envelope of the impulse response and whose temporal resolution Δt should be chosen such that it reflects the temporal resolution ability of the human auditory system. In order to account for frequency dependent absorption the described procedure can be carried out for different frequency bands (typically octave or third octave bands), which yields an energy histogram for each considered frequency band.

It has to be mentioned that in contrast to the image source results these energy histograms can not directly be used for auralization purposes. It is therefore necessary to generate an artificial fine structure for the pressure impulse response based on these time- and frequency dependent energy histograms. This can for example be done by using a random noise process which is then weighted with the square root of the energy envelope of the histograms. A very elegant way is to model the noise process in a way that accounts for the average temporal distribution of reflections in a room impulse response. This can be achieved by using a Dirac impulse sequence based on an inhomogeneous Poisson process[3] which well reflects the average distribution of sound reflections in an "ordinary" room according to equation 3.21. SRT implementations based on such a Poisson sequence were realized by Heinz [1994] or Schröder [2011]. A detailed description of the generation process for such a Poisson sequence is given in Pelzer et al. [2010a]. It is generally considered that in the late part of the IR with a high reflection density and overlap the variations in the room impulse response which are introduced by the stochastic character in the Poisson sequence are generally inaudible to the human ear.

In addition to the stochastic character of the Poisson sequence further simulation inherent parameters like the number of sound particles, the discrete time pattern for collecting the particles or the size of the detection sphere may influence the energy histograms and thus also the resulting impulse response. However, accepted guidelines based on error estimation formulas and reasonable assumptions on the sensitivity of human hearing are given in the literature [Heinz, 1994, Pelzer et al., 2011b], which can be used to minimize the influence of these parameters and which have been followed throughout this thesis. The detailed settings of the simulation input parameters will be given in the application sections 8 and 9 for each simulation.

[3] This means that the probability of having k reflections within a time interval $(t, t+\tau]$ has the following probability density function:

$$P[(N(t+\tau) - N(t)) = k] = \frac{e^{-\mu_{t+\tau,t}}(\mu_{t+\tau,t})^k}{k!}, \quad \text{with} \quad \mu_{t+\tau,t} = \int_t^{t+\tau} \mu(t)dt, \tag{4.10}$$

where $\mu(t)$ is the average number of reflections per unit time at the time instant t. It can be shown that for such a process the time distances between two subsequent reflections are exponentially distributed with expectation $\frac{1}{\mu(t)}$. By setting $\frac{1}{\mu(t)}$ equal to the average distance between two reflections in a room according to equation 3.21 the Poisson sequence gets the desired temporal distribution.

Apart from the above mentioned well controllable uncertainty factors in the SRT results, the applicability of energy based GA simulations (like the SRT) is further limited by different lower frequency bounds. These limitations are mainly due to the negligence of the phase relations in the summation of the incident energy fractions and lead to the following restrictions: Firstly, and most importantly the assumption of a statistical phase is of course only valid at frequencies above the Schroeder frequency f_{S}. A further criterion can be derived from the so-called Waterhouse-effect [Waterhouse, 1955], which describes the observed increase of the sound pressure level in diffuse sound fields at distances closer than one quarter of a wavelength to a room boundary. While this obviously implies that a receiver should be further away from any boundary than $\frac{\lambda}{4}$ in energy based GA simulations a more general criterion is derived by Kuttruff [1998]. He demands that all considered wavelengths should be small compared to the mean-free-path in a room, thus yielding

$$f \gg cS/4V \; . \tag{4.11}$$

A third criterion also considers the receiver position, which should be outside the reverberation radius $r_{\mathrm{H}} = 0.057\sqrt{\frac{V}{T}}$. In this context, again Kuttruff [1991] shows that at a distance of $r \approx 3r_H$ the standard deviation of the phase is already almost 90°, thus implying an almost perfectly random phase beyond this distance.

It is important to mention that these restrictions do generally not apply to the image source method if the complex, time harmonic sound pressure $\underline{p}(\boldsymbol{x})$ is used as the sound field descriptor and the room boundaries are described by their complex and angle dependent reflection factors, since in this case the simulation does account for interference effects. Thus the above mentioned limitations of the SRT can in principle be alleviated by substituting the early part of the impulse response by image source results as is generally done in hybrid geometrical acoustics softwares.

4.2.3. Diffraction models in GA domain

The inclusion of edge diffraction in geometrical acoustics is subject to research work in many places. For first-order diffraction, promising algorithms based on the geometrical Uniform Theory of Diffraction (UTD) are in use. The UTD is a high frequency approximation based on secondary edge sources of diffracted waves which can be reflected and diffracted again on their way to the receiver. Based on this approach Svenson et al. [1999] derived a fast and precise method for the computation of edge diffraction impulse responses which can be included into an image source model. On the other hand Stephenson [2004] derived a very different diffraction model using a deflection angle probability function based on Heisenberg's uncertainty principle. This model is suitable for the inclusion into models based on the dispersion of energy particles such as the ray tracing method.

Recently, implementations and further improvements of such diffraction models in existing or new room acoustic simulation tools have been reported [Schröder and Pohl, 2009]. In the case of the simulation software *RAVEN* the evaluation and validation of these models is currently ongoing and first promising results have been published [Schröder et al., 2011]. Taking into account the computational efficiency of such implementations and the ever-increasing computational processing power, it appears likely that such diffraction models will soon be on the verge of becoming also available in commercial applications. However, to date these diffraction models have yet not found their way into commercial room acoustic simulation tools. In the course of this thesis all geometrical acoustics simulations have therefore been run without accounting for edge diffraction. Taking into account the considered application examples in sections 8 and 9 it has to be admitted that this appears especially problematic in the case of the car passenger compartment where edge diffraction at the seats might be an important factor. However, as we will see in the discussion of the car passenger compartment this is by far not the only problem in this case.

4.2.4. Boundary, source and binaural receiver model in GA domain

Similar to the corresponding section 4.1.2 on FE simulations, the present section describes the boundary, source and receiver models used in room acoustic GA simulations and summarizes the required input data that needs to be obtained for these simulations.

1. **Boundary representation**

   In geometrical acoustics simulations the acoustic behaviour of the room boundaries is generally modeled by assigning the diffuse-field absorption coefficient α_{diff} and scattering coefficient s to each boundary. With regard to the absorption characteristics at the boundary this means that both the phase-shift at the boundary reflection as well as the angle dependence of the reflection characteristics are neglected in typical GA simulations. Although there has been some discussion about the inclusion of the reflection phase (e.g. Suh and Nelson [1999], Jeong et al. [2008]) and the angle dependence of the absorption coefficient [Mechel, 2002] it can be concluded that at least for the late part of the impulse response this simplification is generally admissible. This can be explained by the immense reflection overlap in the late part of the IR (cf. eq. 3.21) which makes it impossible to perfectly reconstruct its temporal fine structure anyway. Only in the early part of the IR which is modeled by the ISM a benefit might be achievable by considering an angle dependent complex reflection factor. This will be investigated in detail in section 7.1. However, it has to be mentioned in advance that a theoretically possible benefit might be blurred by the considerable uncertainty in the determination of the angle dependent complex reflection factor.

   Similar reasoning applies to the scattering coefficient. The major difference here is that in contrast to the absorption coefficient, the scattering coefficient itself already constitutes a huge simplification of the scattering processes at a structured or rough surface since in a strict sense it only applies to ideally diffuse reflecting surfaces. However, a more accurate representation of this scattering process would necessitate the determination of hemispheric scattering patterns for all angles of incidence and all considered frequency bands. Since such a measurement effort is generally infeasible for room acoustic simulations where a large number of surfaces needs to be characterized, the application of the scattering coefficient appears to be an inevitable simplification. Since the scattering coefficient is not in the focus of this thesis, we will however not go any deeper into questions regarding the limitations of the approach, the considerable measurement uncertainty of this parameter or the subjective relevance of differences in the assigned scattering coefficients. Interesting readers are therefore referred to recent work on the topic, which is for example summarized in Vorländer [2010].

2. **Source representation**

   As was already mentioned, sound sources in GA simulations are modeled as infinitesimal point sources from which sound rays are emitted into multiple directions in the room. In order to realistically model the directional and frequency characteristics of a real sound source it is thus necessary to suitably adjust the spherical energy distribution of the emitted sound rays in all considered frequency bands. This is done by applying the so-called directivity function of the considered real sound source which is defined as follows:

   > *The directivity $D(\theta, \varphi)$ of a sound source at an emission angle (θ, φ) is given by the ratio of the free field frequency response of this sound source at angle (θ, φ) and its corresponding on-axis frequency response where both have to be determined at distance r to the source in the far field.*

   The directivity thus gives the directional frequency characteristics of a source relative to its principal direction. Consequently, the absolute frequency characteristics of a source signal can then be accounted for by convolving the simulated impulse response with the source signal measured in the principal direction of the source in an anechoic environment in the far field. This is for example done in the case of musical instruments as sound sources. In the special case of a loudspeaker source this can even be done irrespective of the actual source signal by using the on-axis free field impulse response of the loudspeaker measured at 1 m distance and 1 V at the loudspeaker input.

   Although the point source model with directivity $D(\theta, \phi)$ does in a strict sense not satisfy the wave equation [Mechel, 2002], the described source model consisting of the on-axis free field impulse response of the sound source and its directivity function gives an adequate model of real sound sources if far field conditions apply. The quality of the source model is obviously dependent on the angle and frequency resolution of the directivity data. In GA simulations this is generally given in third octave bands at an equiangular grid of $2°$ to $10°$.

3. **Binaural Receiver representation**

   The auralization of a room acoustic sound field should replicate to the best possible extend the auditory impression that a listener would get if he was present in the room. This implies that in addition to a realistic temporal and spectral envelope of the simulated reverberant sound field the simulation result should also contain directional information, which gives the listener the ability to localize sound sources in the auralized sound field. This information is generally extracted from the characteristic interaural differences between the sound signals at a listener's left and right ear, which are caused by the spatial distance between the two ears as well as by diffraction at the ears, head and torso of the listener. These differences can be described in the frequency domain in the form of the so-called "Head Related Transfer Functions" (HRTF). Blauert [1997] describes three different types of HRTF definitions. Throughout the present thesis only the so-called "free field transfer function" definition will be used which can be summarized as follows:

> *Consider a human listener in the free field whose head is centered at $\boldsymbol{x} = 0$ and a sound source that emits a broadband signal and moves on a sphere with radius r and origin $\boldsymbol{x} = 0$, where r should be chosen sufficiently large for far field conditions to apply. The HRTF for an angle of incidence (θ, φ) is then given by the ratio of the sound pressure at the blocked left and right ear canal entry of the listener and the sound pressure at the position $\boldsymbol{x} = 0$, when the head is not present.*

Coming back to the field of geometrical acoustics simulations, it is clear that when considering a single sound reflection reaching the receiver at an angle of incidence (θ, φ) the inclusion of the respective HRTFs for the left and right ear (by multiplication in frequency domain or by convolution in time domain) generates binaural impulses that account for the diffraction at the listener's head. Thus in order to include the HRTFs in the geometrical acoustics simulation, the angle of incidence of each image source or sound particle has to be monitored during the simulation. In the case of the ISM each image source can then be convolved with its respective HRTF according to its angle of incidence. In the case of the SRT method, a further processing step is necessary since the SRT only gives angle dependent energy histograms, which than have to be transformed into binaural room impulse responses. Details on the necessary processing steps can be found in Schröder [2011] and will not be discussed here. Summing up, it should be noted, that the inclusion of HRTFs in geometrical room acoustic simulations enables the realistic modeling of the directional clues in a simulated room impulse response. However, in a strict sense this is only true if the receiver is in the far field of the considered sound source. Moreover, it should be mentioned that from the point of view of a sound ray or particle the receiver itself is still acoustically transparent. Thus, a redirection of the sound rays caused by reflection or diffraction at the head and torso itself is not taken into account, which might lead to noticeable errors in the case of very small rooms and high frequencies.

4.3. Combination of FE and GA results

Taking into account the reasonable frequency ranges of application for room acoustic FE and GA simulations, the aim of the combination of the two methods is to use the FE results for the modally dominated part and the GA results for the stochastical part of the room transfer function. The combination should thus take place in the frequency domain and the transition frequency $f_{\text{Transition}}$ between both domains has to be chosen according to the lower frequency bounds of geometrical acoustics given in section 4.2.2. However, before it is possible to combine the FE and GA simulation results it is necessary to preprocess the FE results such that they conform to the target signal representation of the GA simulation results, i.e. an impulse response with FFT degree $n \in \mathbb{N}$ and sampling rate 44.1 kHz.

The finite element results are given in the frequency domain for discrete frequency steps Δf in a frequency interval $[f_{\text{Start}}, f_{\text{Stop}}]$. In order to get a conforming signal representation with the GA results with respect to sampling rate and signal length (FFT degree) it is thus necessary to interpolate the FE results to the desired frequency bins of the GA simulations and to pad the transfer function with zeros outside of the interval $[f_{\text{Start}}, f_{\text{Stop}}]$. If the chosen frequency step width of the FE simulation complies with equation 4.9, the interpolation to the GA frequency bin positions is best done by a spline interpolation of the magnitude and unwrapped phase of the frequency domain FE results. From a signal theoretic point of view the frequency domain interpolation could also be conducted by padding zeros at the end of the time domain impulse response. This option is however not readily applicable to the raw FE results, which is mainly due to the rectangular filter with pass band $[f_{\text{Start}}, f_{\text{Stop}}]$ which is inherent in the FE results. When transformed to the time domain, this rectangular bandpass can result in strong acausal artefacts in the corresponding impulse response which complicate the padding of zeros at the end of the IR. In the course of this thesis, we have experimented with many different processing alternatives in order to adjust the signal representation of the FE results to the GA results. As a conclusion it can be said, that if equation 4.9 is observed, the FE results can be successfully processed in different ways both in time and frequency domain. However, the frequency domain spline interpolation of magnitude and unwrapped phase appears to be the easiest option and is therefore recommended here.

Thus, after interpolation to the GA frequency vector, the FE results correspond to a low frequency impulse response that is filtered with a rectangular filter with a pass band of $[f_{\text{Start}}, f_{\text{Stop}}]$. In order to get rid of the non-causalities in the low frequency IR which are caused by this rectangular filter it is recommended to apply a highpass filter at a cutoff frequency $f_{\text{HP,FE}} > f_{\text{Start}}$ with a minimum attenuation of 20 dB at f_{Start} and accordingly a lowpass filter at a cutoff frequency $f_{\text{LP,FE}} < f_{\text{Stop}}$ with a minimum attenuation of 20 dB at f_{Stop}. This way it is assured that the sharp edges of the rectangular filter are sufficiently attenuated. If slight non-causalities remain in the FE impulse response, which can also be caused by inaccuracies in the FE results themselves, these non-causalities can be suppressed by application of a suitable two sided time window, which only extracts the 'valid' part of the IR.

A straightforward approach for combining the IRs from both simulation domains now consists in highpass filtering the GA results at the frequency $f_{\text{HP,GA}} = f_{\text{LP,FE}} = f_{\text{Transition}}$, and then simply adding the filtered frequency responses. Taking into account that the phase relation of both simulation methods can be regarded as stochastic in the cross-fade frequency range, filters with 3 dB attenuation at the cut-off frequency should be used for the combination. Moreover, since it was found that using ordinary Butterworth filters causes a frequency dependent latency in the impulse response (i.e. the low frequency part of the IR is delayed by a few more milliseconds than the high frequency part), zerophase filters can be used to avoid this effect. This is of course bought at the price of a slight preringing of the acausal zerophase filters.

It should be mentioned at this point, that independent of the chosen filter type (zerophase or Butterworth) no audible artefacts could be heard in auralizations based on the combined impulse responses. It was therefore however not possible to favor the one or the other based on a subjective comparison of the auralization files. Apart from the choice of the filter type, a slight disadvantage of this combination technique is that it requires FEM calculations up to frequencies approximately one third octave band higher than the transition frequency in order to prevent an abrupt step in the filtered FE frequency response. Although this is rather costly in terms of computation time, Granier et al. [1996] and Nöthen [2008] report that approaches which only calculate FEM results up to $f_{\text{Transition}}$ and then try to 'stitch' the results to the GA-based results generally seem to suffer from visible and audible artifacts. Figure 4.1 summarizes the discussed processing steps of the presented combination method in a block diagram.

Further information on possible combination methods for FE and GA simulation results in the frequency domain are reported in the master's thesis by Nöthen [2008]. The above described method accounts for the results and recommendations given in this master's thesis. As mentioned above, the combination method has been evaluated by subjective listening tests and no audible artefacts have been reported for combined impulse responses which were generated according to this specification.

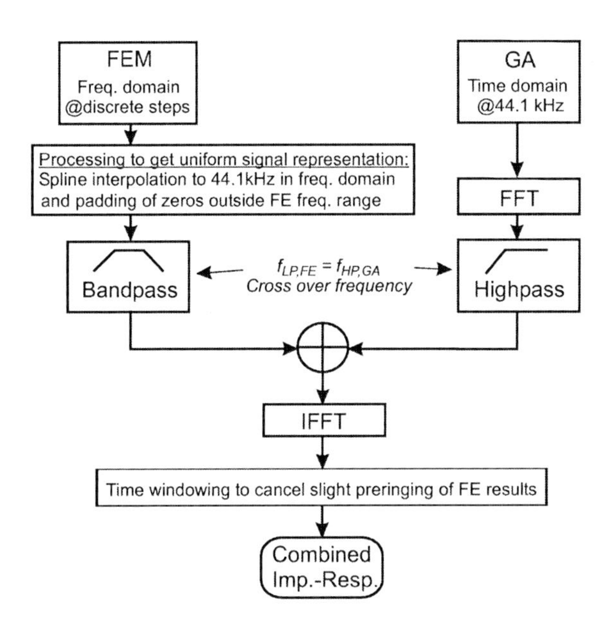

Figure 4.1.: Block diagram of combination method for FE and GA simulation results.

5. Determination methods for the required boundary conditions

The current chapter introduces the measurement and calculation methods that have been applied to determine all necessary boundary condition data for the combined FE/GA simulations which will be presented in sections 7, 8 and 9 of this thesis. As discussed in section 3.2.1 the acoustic reflection characteristics of a wall or absorber surface can in many cases be characterized in good approximation by its acoustic surface impedance $\underline{Z}_S$ or in the case of purely energetic considerations by its absorption coefficient α. In section 4 we have then shown how these quantities can be applied as acoustic boundary conditions in room acoustic FE and GA simulations respectively. While sections 3.2.1 and 4 have focused on the definition of these quantities and their application as boundary conditions in RA simulations, the current section discusses the theoretic background of the corresponding measurement and calculation methods with regard to their limitations and possible application problems.

Over the past decades various methods for the determination of the acoustic surface impedance $\underline{Z}_S$ and the absorption coefficient α have been proposed in literature and some of them have even been standardized in corresponding ISO standards. Without claiming to be exhaustive, table 5.1 gives an overview of the most commonly cited methods with references to the respective publications or standards. A nice summary of the working principles of these methods is given in the master's thesis by Praast [2009]. In the course of the present thesis only four of these methods (marked with an asterisk in table 5.1) have been closely investigated. The chosen methods give a representative cross-section of existing laboratory, in-situ and analytic methods and will be discussed in the following subsections. In some cases modifications to the presented methods are suggested, that were found to improve the overall quality of the results. A comparison of these methods based on the measurement results obtained for the same material configurations and a comparative discussion of their strengths and weaknesses will then finally be given in the context of the application examples in sections 8 and 9.

Table 5.1.: Overview of procedures for the determination of the reflection characteristics of absorbing materials.

	Standardized Procedures		Alternative Laboratory Approaches		
Method	Impedance Tube*	Reverberation Room*	Barry Method	Cepstrum Method	Allard Method
Publication	ISO 10534-2:1998	ISO 354:2003	Barry [1974]	Bolton and Gold [1984]	Allard et al.[2]
Applicable freq. Range	100 Hz to $f_{\text{cut-on}}$ [1]	$f > (1000/\sqrt[3]{V})$ Hz	$f > 500$ Hz	100 Hz to 10 kHz	$f > 500$ Hz
Sound incidence	normal inc.	diffuse inc.	normal inc.	angular inc.	angular inc.
Sound field model†	PW	DF	MS	MS	SR
Measurement quantity	p-mic (min. 2 pos.)	p-mic (multiple pos.)	2 p-mics	1 p-mic	1 pp-probe
Target quantity	$\underline{Z}_{\text{S,norm}}$	α_{diff}	$\underline{Z}_{\text{S,norm}}$	$\underline{Z}_{\text{S}}(\theta_{inc})$	$\underline{Z}_{S,norm}$
Ref. meas. required	no	no	yes	yes	no

	Alternative In-Situ Approaches				Modelling Approach
Method	Microflown Method*	Subtraction Method	Nocke Method	EA-Noise Method	2-Port Network Model*
Publication	de Bree et al.[3]	Mommertz [1995]	Nocke [2000]	Takahashi et al. [2005]	Mechel [1998, pp.47ff]
Applicable freq. Range	100 Hz to 20 kHz	250 Hz to 8 kHz	80 Hz to 4 kHz	$f > 200$ Hz	20 Hz to 20 kHz
Sound incidence	angular inc.	angular inc.	angular inc.	diffuse inc.	angular inc.
Sound field model†	PW, MS or SR	PW	SR	DF	PW
Measurement quantity	1 pu-probe	1 p-mic	1 p-mic	1 pp- or pu-probe	material param. of layers
Target quantity	$\underline{Z}_{\text{S}}(\theta_{\text{inc}})$	$\underline{Z}_{\text{S}}(\theta_{\text{inc}})$	$\underline{Z}_{\text{S}}(\theta_{\text{inc}})$	$\underline{Z}_{\text{S,diff}}$	$\underline{Z}_{\text{S}}(\theta_{\text{inc}})$
Ref. meas. required	yes	yes	yes	no	-

(*) In the course of this thesis only the procedures marked with an asterisk will be further discussed.

(†) Abbreviations for the sound field models: PW = plane wave; DF = diffuse field; MS = mirror source, i.e. the original point source is mirrored at the sample surface; SR = spherical reflection (cf. section 3.2.3).

(1) $f_{\text{cut-on}}$ marks the frequency where the first lateral eigenmodes become propagable in the tube and thus the plane wave assumption for the calculation of the reflection factor is no longer fulfilled.

(2) Jean F. Allard and his co-workers have published various articles on the topic of surface impedance measurements based on the pressure measurement of two microphones in close vicinity to the measurement samples. A selection of these papers is given in the following: Allard and Sieben [1985], Champoux et al. [1988], Allard et al. [1989, 1992].

(3) Various authors have applied the Microflown pu-probe to the measurement of the acoustic surface impedance. Some key papers on the method and the pu sensor are de Bree [2003], Lanoye et al. [2006], Alvarez and Jacobsen [2008]. A nice and extensive overview on the probe, the impedance measurement and the relevant literature is also given in the Microflown E-Book [de Bree, 2009]. However, it should be kept in mind that this book is published by the manufacturer of the probe.

5.1. Impedance tube: Transfer function method

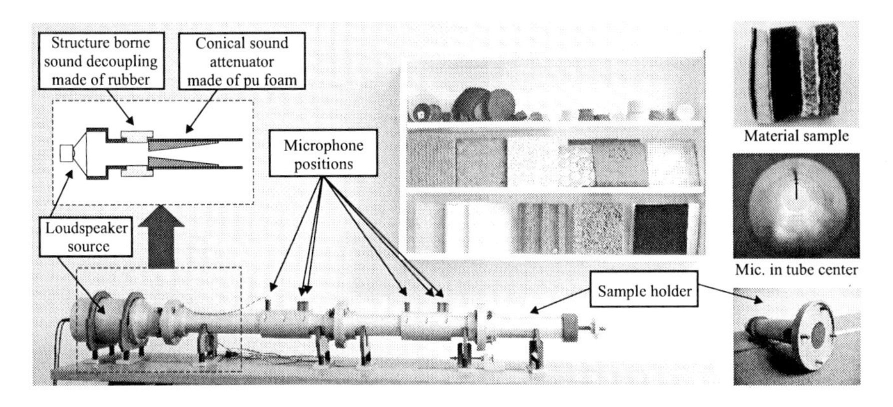

Figure 5.1.: The impedance tube at ITA of RWTH Aachen University.

The impedance tube transfer-function method is described in detail in ISO 10534-2:1998, "Acoustics - Determination of sound absorption coefficient and impedance in impedance tubes - Part 2: Transfer-function method". Figure 5.1 shows the cylindrical impedance tube at ITA of RWTH Aachen University with its various microphone slots and the source loudspeaker on the left end and the material sample on the right end of the tube. The measurement technique is based on the fact that assuming plane wave propagation in the tube the complex reflection factor for normal sound incidence can be extracted from the transfer function $\underline{H}_{12}$ between two microphones in front of the material sample under consideration. The corresponding formula is given as follows:

$$\underline{R} = \exp(j2ks)\frac{\exp(jkd_{12}) - \underline{H}_{12}}{\underline{H}_{12} - \exp(jkd_{12})}, \tag{5.1}$$

where s is the distance between the reference plane of the absorber and the closest microphone and d_{12} is the distance between the two microphones. Obviously an upper frequency bound of the method stems from the plane wave assumption underlying the calculation of the reflection factor from the transfer function $\underline{H}_{12}$. In a strict sense the plane wave assumption only holds for frequencies below which no lateral eigenmodes can be excited in the tube. Thus, according to the ISO standard the results in a cylindrical tube (as was used in the present thesis) can only be evaluated up to a frequency of $f_{\text{cut-on}} = \frac{0.58c}{D}$, where c is the speed of sound, D is the diameter of the tube and $f_{\text{cut-on}}$ corresponds to the eigenfrequency of the lowest lateral eigenmode with $(n_\text{r}, n_\varphi, n_\text{z}) = (0, 1, 0)$.

On the other hand the lower frequency bound of the method is mainly determined by two factors, which are given by the lower frequency limit of sound radiation of the used loudspeaker and the spacing of the two microphones d_{12}. While the former problem can be dealt with by choosing a suitable loudspeaker which generates sufficient sound power at low frequencies, the latter deserves a little more attention. It can be shown that by using the transfer function method according to equation 5.1 the reflection factor can not be evaluated at frequencies where the spacing of the microphones corresponds to multiples of $\frac{\lambda}{2}$, which is due to the fact that both the nominator and denominator in equation 5.1 tend to zero at these frequencies, which makes the whole term numerically instable. Thus, in order to avoid errors at these frequencies the microphone spacing should be chosen smaller than half the wavelength $\lambda_{\text{cut-on}}$ which corresponds to the highest considered frequency $f_{\text{cut-on}}$. On the other hand such a small spacing is disadvantageous for very low frequencies since due to the long wavelength the differences between the two measurement positions become very small and thus the result is very sensitive to noise or slight inaccuracies in the measurement setup.

Further issues which are often encountered in an impedance tube are related to the mounting of the material samples inside the tube. These problems often occur in the case of a too small or too large cut of the material samples where errors are caused by leakage between the tube boundary and the absorber sample or by clamping effects when the absorber is compressed in the tube. Finally, the formation of eigenmodes in the longitudinal direction of the tube can cause measurement errors at the corresponding eigenfrequencies which is especially problematic in the case of low absorbing material samples.

The following paragraph therefore gives some design guidelines for an impedance tube setup which helps to overcome or at least alleviate some of the above discussed problems. In particular, the setup extends the valid frequency range of the results in both directions, suppresses the formation of z-eigenmodes in the tube and alleviates the problems related to the mounting of the material samples. Some of the presented design guidelines have already been realized and tested on the ITA impedance tube and are therefore marked with asterisk.

- (*) The first modification concerns the upper frequency bound f_c of the tube. According to the ISO standard the upper frequency bound of the ITA impedance tube is given at approximately $f_c = 3920\,\text{Hz}$. By placing the microphones in the centre of the tube, as done in the ITA tube, and not flush at the boundary of the tube as suggested in the ISO standard, the frequency limit f_c can be pushed further to higher frequencies by canceling out the contributions of all lateral modes with $n_\phi \geq 1$, since these modes have a pressure node at the centre of the tube. This way the results obtained in the impedance tube at ITA can be evaluated up to the eigenfrequency of the first radial eigenmode $(n_r, n_\phi, n_z) = (1, 0, 0)$ at approximately 8.16 kHz, which corresponds to the first lateral eigenmode with non-zero contributions at the tube centre.
- (*) The second modification aims at resolving the dilemma regarding the microphone spacing d_{12}. In order to avoid this earlier mentioned problem the measurement method has been extended to the use of three differently spaced pairs of microphone positions. The results of the different microphone pairs are then cross faded in the frequency domain, such that d_{12} is chosen to provide a sufficient phase shift between the two microphones but at the same time $d_{12} < \lambda/2$ is fulfilled for all frequencies below 8.16 kHz.

- (*) As a third measure, a special conical sound attenuator was inserted in the tube directly in front of the loudspeaker in order to dampen the formation of the z-eigenmodes as much as possible. In order to account for the strong high frequency attenuation of the loudspeaker signal by this absorber, the excitation sweep is amplified by a suitable high shelving filter. Alternatively, it would possibly be even more effective to mount the loudspeaker on the side of the tube and use an anechoic end duct at the front side of the tube to cancel the z-eigenmodes. This was however not tried on the ITA impedance tube so far.
- An alleviation of the problems related to leakage or clamping effects can possibly be achieved by the use of a much larger tube diameter, since in a larger tube the edge effects play a less important role. Moreover, the use of a tube with a square instead of a circular cross-section would also facilitate the precise cutting of absorber samples. While the use of a larger tube generally decreases the upper frequency limit of the method, suitable multiple microphone techniques can be applied to considerably extend the valid frequency range by canceling the lowest lateral eigenmodes; e.g. in the case of a quadratic tube with side length a where the tube cross-section is centered at $(x, y) = (0, 0)$ the sum of four calibrated microphones mounted at $P_{1,\dots,4} = (\pm\frac{a}{4}, \pm\frac{a}{4})$ cancels all contributions from lateral eigenmodes with $n_x, n_y \leq 2$.

5.2. In-situ method using a Microflown pu-probe[1]

In the course of this study a Microflown Mini 1/2" pu-probe was used, which consists of a miniature pressure microphone (Knowles FG series) and a Microflown acoustical particle velocity sensor mounted in close vicinity to each other ($\approx 1\,\mathrm{mm}$) in a compact packaging as shown in Figure 5.2. The Microflown particle velocity sensor is based on the principle that the air fluctuations in a traveling acoustic wave cause a temperature difference between two closely spaced heated wires resulting in an electric resistivity difference which linearly relates to the particle velocity. Details on the working principle of the Microflown sensor can for example be found in de Bree et al. [1996] and de Bree [2003]. The outputs of the p and u channel of the sensor are processed in a special Microflown signal conditioner (MFSC-2) and then fed into an AD converter for further processing of the raw measurement data in *MATLAB*.

Provided the case that the sensors and the measurement chain are free field calibrated to give absolute levels of sound pressure in Pa and particle velocity in m/s a simple subdivision of the calibrated input channels yields the local field impedance in the measurement spot. However, a precise free field calibration of the magnitude and phase response of the pu sensor is not easy and different methods have been proposed in the literature. Most of these methods are based on the comparison of the pu sensor output with that of a calibrated reference pressure microphone under well-controlled sound field conditions where the acoustic field impedance $\underline{Z}_\mathrm{f}$ is known. A nice overview of the different applicable calibration techniques is given in de Bree [2009]. In order to determine the acoustic

[1] The following description of the working principle of the Microflown pu sensor is in large parts taken from Aretz et al. [2010a]. Please note, that for better readability, no quotation marks are used to mark identical paragraphs.

Figure 5.2.: Microflown Mini 1/2" pu-probe and microscopic picture of the wires in the sensor, taken from Jacobsen and de Bree [2005].

impedance at the surface of the measured absorber it is in a next step necessary to carry out a suitable transformation of the field impedance $\underline{Z}_{\mathrm{f}}$ measured at a distance h from the sample to the absorber surface. Obviously the transformation rule to apply depends on the assumptions made for the sound field that is incident on the absorber and suitable simplifications need to be made to keep the complexity of the sound field model at a manageable level. Alvarez and Jacobsen [2008] describe three different models that can be used for this transformation. For brevity, only a short summary of the basic assumptions of these models is given in this thesis. For more details and the according formulas the reader is referred to the original publication by Alvarez and Jacobsen [2008]. The simplest model assumes plane waves, thus the impedance at the absorber surface $\underline{Z}_{\mathrm{S}}$ is given by simple transformation of the measured field impedance $\underline{Z}_{\mathrm{f}}$ over a 1-dimensional transmission line of length h. The second model assumes an ideal point source and a corresponding image source which is mirrored on the absorber surface and whose amplitude is attenuated by the complex plane wave reflection factor $\underline{R}$ of the measured absorber. In contrast to the first model this model accounts for the phase and amplitude shift that is caused by the different distance of the true and the image source to the measurement probe. Finally, the third and most complex model also assumes an ideal point source for the incident wave but for the reflected wave it accounts for the fact that due to the angle dependency of the reflection factor the reflected wave can in a strict sense not be fully described by a simple image source monopole anymore (cf. section 3.2.3 for more details). This phenomenon can be accounted for by using the so-called spherical reflection coefficient. Unfortunately the relation between the measured field impedance and the sought-after surface impedance can in this case not be given in closed form and therefore the surface impedance has to be determined by a computationally expensive iterative method.

Although the third model has to be considered the most realistic sound field model for the impedance measurement setup, Alvarez shows in his paper that the differences between the 2nd and 3rd model are very subtle, especially when the measurement is carried out very close (5 mm) to the specimen. Moreover, in a real measurement situation he was not able to deduce more reliable measurement results from using the 3rd model compared to the 2nd one.

The measurement of the acoustic surface impedance therefore consists of three major steps. Firstly, the pu-sensor has to be carefully calibrated. Secondly, the local field impedance at a defined distance to a flat absorber surface has to be measured with the calibrated pu-probe and finally, this field impedance has to be transformed to the absorber surface to get the desired surface impedance, from which the plane wave reflection coefficient $\underline{R}$ and the absorption coefficient α of the absorber can be determined.

In the course of this doctoral thesis various studies and measurement series have been conducted to better quantify the uncertainty factors in the in-situ measurement of acoustic surface impedances with the Microflown pu-probe. The essence of these studies is summarized in the discussion of the measurement results obtained for the studio and car materials in sections 8 and 9. A more extensive elaboration on the uncertainty factors of the Microflown impedance measurement setup can be found in the diploma thesis by van Gemmeren [2011]. The very interesting results of this diploma thesis shall be published in a suitable journal in due time.

5.3. Two-port network model for layered absorbers

In the course of this thesis a two-port network model for the calculation of the acoustic surface impedance of layered absorber configurations was implemented in *MATLAB* following the specifications given by Mechel [1998, p.47ff]. The model assumes plane wave incidence on a laterally extended layered absorber, where each absorber layer has a constant thickness and homogeneous material properties. In this case the mathematical description of the sound propagation inside the layered absorber can be reduced to a one-dimensional model, where the absorber is modeled as a cascade of two-port networks representing the single layers of the absorber. Each absorber layer thus needs to be modeled by a suitable two-port network. Depending on the extent of the phase shift of the sound wave when traveling through the absorber layer this can be done by either using a transmission line model or an equivalent network consisting of concentrated elements. This question obviously relates to the thickness of the absorber layer, the considered frequency range and the speed of sound in the layer. As a rule of thumb, thick porous absorber layers or air gaps generally require a transmission line model, while thin plates, foils or fabric coverings are mostly modeled with concentrated elements. Table 5.2 gives an overview of all absorber layer models that have been implemented in the software tool with reference to the respective original publications and a list of the input parameters for each model.

Before we can calculate the input impedance of this cascaded two-port network it is finally necessary to specify a termination for the absorber, which is generally done by using either a rigid (open-circuit, $\underline{Z}_\mathrm{L} \to \infty$) or free field ($\underline{Z}_\mathrm{L} = \rho_0 c$) termination. The acoustic surface impedance of the layered absorber can then be determined straight forward by calculating the ABCD parameters[2] [Ghosh, 2005, p.353] for each two-port network (i.e. absorber layer) and cascading the two-ports by simple matrix multiplication. Figure 5.3 shows an example of how an absorber can be modeled as a cascade of two-port networks. Conse-

[2]The ABCD-parameters are also commonly known as chain, cascade, or transmission line parameters.

quently, the calculation model allows the determination of the acoustic surface impedance of an arbitrary configuration of absorber layers. Moreover by application of Snell's law of refraction and an according adaption of the propagation constants and characteristic impedances in the absorber layers, the model can also account for angular incidence on a laterally extended layered absorber configuration.

Although the presented two-port network approach and the underlying models for porous layers, foils, air gaps, perforated plates or fabric covers have their obvious limitations, an important advantage of this approach is that it supplies reasonable surface impedances and absorption values throughout the whole audible frequency range, which is generally not the case for the presented measurement techniques. Moreover, since the model allows the calculation of the surface impedance and absorption coefficient for angular incidence, it also enables the calculation of the diffuse or field incidence values of these quantities, which are commonly used in room acoustic simulations.

Table 5.2.: Summary of implemented absorber layer models for two-port network model according to Mechel [1998].

Layer Models	**Parameters**[†]	**Reference**
Fluid models		
Air	T, ϕ, p_0	
Porous Absorber Models		
Zwikker Kosten model	Ξ, σ_v, χ, κ_{eff}, ρ_a [‡]	Zwikker and Kosten [1949], Mechel [1995, p.85ff]
Delanz Bazley model	Ξ	Delany and Bazley [1970]
Miki model	Ξ	Miki [1990]
Komatsu model	Ξ	Komatsu [2008]
Plate models		
surface mass	ρ_m	Mechel [1995, p.771]
mass with flow resistivity	ρ_m, Ξ	Mechel [1995, p.772]
elastic plate*	ρ_m, E, ν, η	Mechel [1998, p.307]
micro perforated panel**(MPP)	ρ_m, σ_s, d	Maa [1998]
perforated panel***	σ_s, d	Mechel [1995, p.655ff], Mechel [1998, p.55-56]

Parameter legend: T = temperature [°C]; ϕ = humidity [-]; p_0 = static ambient pressure [Pa]; Ξ = static flow resistivity [Pa s/m²]; σ_v = volume porosity of porous absorber [-]; χ = structure factor [-]; κ_{eff} = adiabatic exponent of air, corrected for effects due to thermal relaxation in the absorber [-]; ρ_a = absorber density [kg/m³]; ρ_m = density of plate [kg/m³]; E = Young's modulus [Pa]; ν = Poisson's ratio [-]; η = plate loss factor [-]; σ_s = perforation ratio (surface porosity) [-]; d = hole diameter [m].

† For all models the thickness t of the layer needs to be specified.

‡ In the classical Zwikker Kosten model the absorber density is not included. The inclusion of the density is implemented according to Mechel [1995, p.98] and accounts for the fact that at low frequencies the whole absorber frame behaves like a mass that moves in phase with the air particles in the absorber which impedes the friction between the air particles and the absorber frame.

* With consideration of coincidence effect.

** With optional consideration of coincidence effect, in this case parameters of elastic plate have to be added.

*** In contrast to the MPP model this model does not account for damping due to friction at the interior walls of the holes and at the edges of the holes. This is admissible if the hole diameter is large compared to the thickness of the plate.

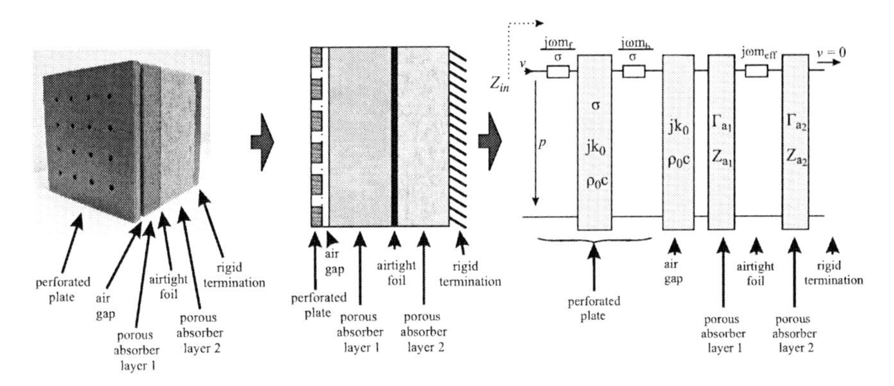

Figure 5.3.: Schematic diagram of two-port network model for layered absorber configurations.

5.4. Sound absorption measurement in a reverberation room

The measurement of the diffuse field absorption coefficient α_{diff} of a material sample in a reverberation room is described in detail in ISO 354:2003 "Acoustics - Measurement of sound absorption in a reverberation room". The measurement procedure is based on the evaluation of the differences in reverberation time between the empty reverberation chamber and that of the reverberation chamber with material sample in it. Under the assumption of a diffuse sound field in both cases the third octave band diffuse field absorption coefficient of the sample under consideration can be deduced by using the Eyring or Sabine reverberation law. The results are generally averaged over multiple source and receiver positions and possibly also different positions of the absorber sample in the room.

The literature on sound absorption measurements in a reverberation room is extensive and can be roughly categorized into contributions dealing with:

- Edge effects at the absorber boundaries
  [Esche, 1967, Dekker, 1974]
- Influence of sound diffuser panels on the results
  [Benedetto et al., 1981, Kuttruff, 1981]
- Description of the state of diffusion in the room
  [Schultz, 1971, Nelisse and Nicolas, 1997]
- Low frequency extension of the validity of the results
  [Davern, 1987, Zha et al., 1999].

Further important contributions which deal with several of the above mentioned aspects were for example published by [Kuhl, 1983, Cops et al., 1995].

Figure 5.4.: The reverberation room at ITA of RWTH Aachen University with diffusing panels and absorber test sample on floor.

While the diffraction effects at the absorber boundaries generally lead to overestimated absorption values, the insufficient diffusity of the sound field in the room with absorber sample generally leads to underestimated absorption values. Guidelines to alleviate these problems are also prescribed in the ISO standard. In particular, it is required that the absorber sample is enframed by a rigid boundary and the reverberation chamber should be designed without parallel walls and with a sufficient number of randomly distributed low-absorptive diffusor panels in the room or on the room walls.

Another problem of this method is that at low frequencies the variation of measurement results with regard to the source and receiver position considerably increases due to an insufficient modal density. While this problem can be alleviated by sufficient averaging, systematic errors might prevail in the lower frequency bands. These errors occur if the low frequency sound decay is dominated by eigenmodes with contributions in only one or two room dimensions, which can occur in rooms with parallel walls and insufficient low frequency diffusion. In this case the application of the Eyring or Sabine reverberation law is generally not admissible since these modes generally have a longer decay time as the three dimensional modes. Kuttruff [2000] gives an experimental lower frequency bound of the reverberation room method as $f_{\mathrm{g}} > \frac{1000}{\sqrt[3]{V}}$ Hz, where V is the room volume. According to the ISO standard measurements can be evaluated down to the 100 Hz third-octave band, which roughly corresponds to the limit given by Kuttruff for rooms greater than $200\,\mathrm{m}^3$. It should be mentioned that this experimental lower frequency bound is surprisingly far below the Schroeder frequency of typical reverberation rooms built to the requirements of the standard ($f_S \approx 300 - 400\,\mathrm{Hz}$). This is possibly due to the fact that (a) as already mentioned the use of multiple source and receiver positions partly cancels errors caused by insufficient modal density and (b) systematic errors caused by low frequency eigenmodes with strongly deviant decay times compared to the Eyring or Sabine prediction will partly cancel out in the calculation of the diffuse field absorption coefficient if the deviations are similar in the measurement with and without absorber in the room.

Unfortunately the measurement method provides no means for determining the phase of the reflection factor and thus the acoustic surface impedance can not be determined by this method. Consequently, this method is generally not suitable for the determination of boundary conditions for room acoustic FE simulations. Only in the case of relatively hard and reflective walls it is admissible to assume $\arg(R) = 0$, which enables the calculation of a real valued surface impedance from the measured absorption coefficient.

Since the absorption coefficient is measured for diffuse incidence the calculation of a real valued, locally reacting surface impedance has to be carried out based on the inversion of the following formula:

$$\alpha_{\text{diff}} = 8\frac{1}{z}\left(1 - \frac{1}{z} \cdot \ln(1 + 2z + z^2) + \frac{1}{1+z}\right), \quad \text{with} \quad z = \frac{\underline{Z}_{\text{S}}}{Z_0}. \tag{5.2}$$

This formula is derived from a formula given by Mechel [1995, p.15][3] where we let the imaginary part of the impedance tend to zero. In earlier studies [Aretz, 2009] we have applied a simpler approach, i.e. to calculate the real-valued acoustic surface impedance from the measured absorption coefficients as $\underline{Z}_{\text{S}}^{*} = Z_0 \frac{1+R}{1-R}$, with $R = \sqrt{1 - \alpha_{\text{diff}}}$. This approach however introduces a considerable overestimation of the damping of the room walls since it implicitly assumes that $\alpha_{\text{diff}} = \alpha(0)$. In the case of hard and reflective materials this can lead to an error in the impedance magnitude of almost 50% (between $\underline{Z}_{\text{S}}$ calculated from eq. 5.2 and $\underline{Z}_{\text{S}}^{*}$). However, also the improved approach given in equation 5.2 can only give reasonable results in cases where the reflection factor phase can be neglected in good approximation. It is therefore emphasized again that the deduction of a surface impedance with a considerable imaginary part is not possible based on a measured diffuse field absorption coefficient.

In the course of this thesis all reverberation chamber absorption measurements have been carried out in the reverberation chamber at ITA of RWTH Aachen University which is shown in figure 5.4. The measurement method has mainly been applied to boundary surfaces with a very inhomogeneous shape or structure that could not be satisfactorily characterized by any of the other three presented methods. Further reasons for the inapplicability of the impedance tube, Microflown or two-port network model will be given in the applications sections 8 and 9, where the measurement of the specific boundary materials is discussed in detail.

[3]The original reference is Mechel [1989, p.72] but the formula contains an error here. The corrected formula is given in the errata section of the second volume of Mechel's books, which is referenced above.

6. Determination methods for the source and receiver conditions

6.1. Sound source characterization

In the course of this thesis we are mainly concerned with the modeling of the sound emission from loudspeaker sound sources. Although the modeling of other sound sources like musical instruments or human speakers is not explicitly a subject of this thesis, the following considerations regarding the source characterization similarly apply to these sources. This is due to the fact that in room acoustic GA simulations it is common practice to model any sound source by its source directivity and an anechoic far-field recording of a sound signal which is measured along the reference axis of the sound source. As was mentioned earlier, the main difference between loudspeakers and other sources is however, that a loudspeaker can be modeled irrespective of the actual audio signal that is fed into it by considering its free field transfer function in the main direction of radiation, which can then be convolved with arbitrary audio material. The present section discusses how the necessary loudspeaker input data for combined FE/GA simulations was determined in the course of this thesis and establishes some links between the different methods.

6.1.1. Free field pressure, membrane velocity and directivity measurement

As we have discussed in section 4.2.4 a sound source in the GA simulation domain is described by its pressure frequency response and directivity function measured in the far-field in an anechoic environment. Detailed guidelines to the measurement of these quantities are given in the standard "IEC 60268 – 5:2003 + A1:2007: Sound system equipment – Part 5: Loudspeakers"[1] and shall not be repeated here. Following these guidelines three different measurement setups have been used in the course of this thesis to determine the loudspeaker free field response and directivity function in the hemianechoic chamber at ITA of RWTH Aachen University. Firstly, measurements of loudspeaker drivers without chassis were carried out with the loudspeaker mounted flush into the hard reflective floor of the hemianechoic chamber with a large heavily damped cavity behind it. By taking advantage of the radial symmetry of most loudspeakers drivers the directivity can in this case be determined by measuring the pressure responses on a quarter circle as shown in figure 6.1 (a). Secondly, loudspeakers with speaker cabinet were measured on a special

[1]The German version of the standard is: DIN EN 60268 – 5:2003 + A1:2009: Elektroakustische Geräte – Teil 5: Lautsprecher.

swivel arm which is installed in the hemianechoic chamber at ITA with the microphone placed on the floor at 8 m distance on the reference axis of the loudspeaker as shown in figure 6.1 (b). Thirdly, a special measurement setup where the loudspeaker is mounted on a turntable and a swivel arm with a microphone rotates around the loudspeaker was used. This setup is shown in figure 6.1 (c).

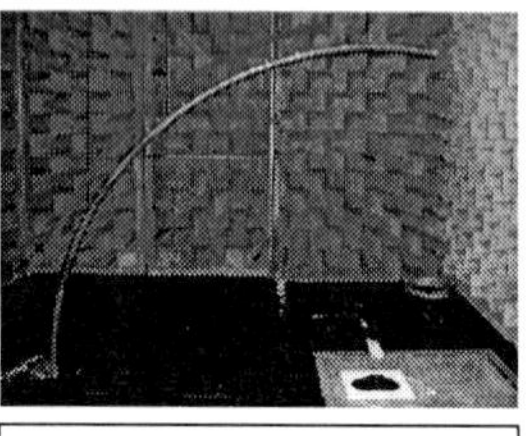

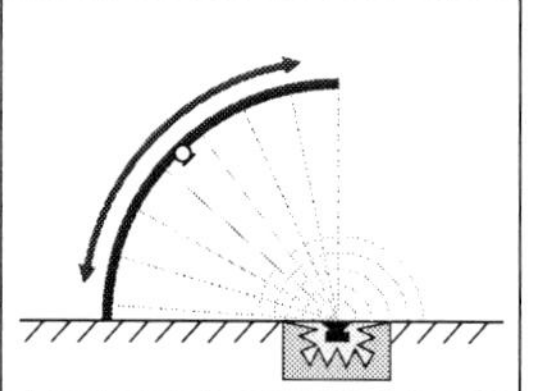

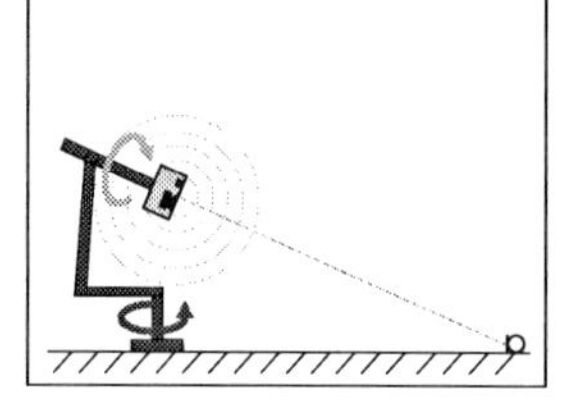

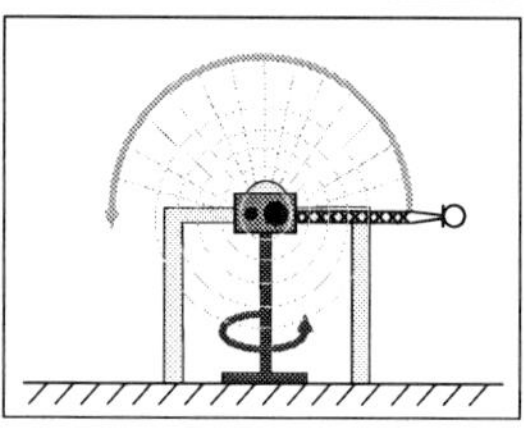

(a) loudspeaker mounted flush in floor, microphone on 90° arch

- Determination of single driver directivity irrespective of diffraction on loudspeaker box or interference with other drivers
- Only applicable if loudspeaker is rotation symmetric
- Bass response of free field pressure transfer function might change when loudspeaker is mounted in chassis due to mechanical compliance of the box cavity.
- Microphone can be moved freely on the arch by hand, generally 5° resolution
- View vector of loudspeaker points to north pole

(b) loudspeaker on swivel arm, mic. on floor

- Full sphere can be measured
- No time window necessary with microphone on the floor
- Possible errors due to diffraction at swivel arm for measurement of rear hemisphere (might cancel with averaging in 3rd octave bands)
- Rigid construction can be used for large and heavy loudspeakers
- Fully automatic measurement, resolution freely adjustable by step motor settings.
- Source receiver distance 8 m
- View vector of loudspeaker points to north pole, therefore high resolution in main direction of radiation

(c) loudspeaker on turn table, mic. on swivel arm

- Only upper half sphere can be measured
- Make use of loudspeaker symmetry to get full sphere (e.g. put loudspeaker on the side as shown in the figure) or turn loudspeaker upside down to measure lower half sphere
- Floor reflection must be cancelled by suitable time window. This might cause problems at low frequencies (<200 Hz), but in this range most loudspeakers are non-directional anyway
- Fully automatic measurement, resolution freely adjustable by step motor settings.
- Positioning uncertainty of swivel arm approximately 1 degree.
- View vector of loudspeaker points to azimuthal plane (poles on top/bottom or on loudspeaker sides). Thus lower resolution in principal direction of radiation

Figure 6.1.: Different measurement setups for the measurement of the loudspeaker directivity and on-axis free field pressure transfer function.

In addition to the free field pressure response and directivity data, in some cases also the membrane velocity of the considered loudspeaker was measured at 1 V input voltage to obtain the necessary source data for the room acoustic FE simulations. The membrane velocity was measured using a Laser Doppler Vibrometer. In the case of bass reflex loudspeakers, where the low frequency sound is also radiated from the open bass reflex holes a laser measurement is however not possible. While it would still be possible to measure the velocity in the holes with for example a Microflown velocity sensor (as presented in section 5.2) a much easier approach can be applied under the reasonable assumption that the loudspeaker has an almost omnidirectional directivity at low frequencies. This approach consists of calculating the equivalent volume velocity $Q(f)$ that an ideal point source would need to generate the measured free field pressure frequency response of the considered loudspeaker, which contains the contributions from the bass reflex holes and all active membranes. This approach of extracting an FE source representation from the commonly used free field pressure transfer function $\underline{p}(f)$ of a loudspeaker is very useful and the corresponding formula that links the free field pressure response to the volume velocity is given as follows:

$$\underline{Q}(f) = \underline{p}(f) \cdot \frac{2r}{j\rho_0 f}, \tag{6.1}$$

where r is the distance at which the free field pressure TF was measured and $\rho_0 = 1.205\,\mathrm{kg/m^3}$ is the density of air. Generally $\underline{p}(f)$ is given at 1 m distance and 1 V or 1 W at the loudspeaker input. However, it has to be considered that this simplification is only admissible in the frequency range where the directivity of the sound source can be considered in good approximation as omnidirectional. If this is not the case, the actual directivity of the sound source can always be considered in the FE domain by a full loudspeaker model comprising all radiating surfaces with their respective velocities and the loudspeaker cabinet.

Similarly, it is possible to determine an equivalent membrane velocity for a loudspeaker that is mounted flush into a wall by using the theoretical relation between sound pressure and membrane velocity for a piston source in a rigid baffle with infinite lateral dimensions, which yields:

$$\underline{v}_{\text{piston}}(f) = \underline{p}(f) \left(\rho_0 c \cdot (e^{-jkr} - e^{-jk\sqrt{r^2+a^2}})\right)^{-1}, \tag{6.2}$$

where again r is the distance at which the free field pressure TF was measured and a is the piston diameter, which can be set to the effective membrane surface of the considered loudspeaker unit.

6.1.2. Electrical analog network model of a loudspeaker

As an alternative to the direct measurement of the loudspeaker free field pressure and membrane velocity, it is possible to use an electrical analog network to model the low frequency performance of a loudspeaker system and thus determine its membrane velocity and radiated sound power as a function of the input voltage at the voice-coil terminals. Such an electrical analog network can be constructed by considering the loudspeaker system as a coupled circuit consisting of an electrical, a mechanical and an acoustical part, where each part can be modeled as a system of idealized lumped components and the parts are interconnected by suitable coupling terms. The equivalent circuit of a loudspeaker in a vented box is given in figure 6.2 with a legend explaining all its components. It can be seen that the electrical part describes the voltage current relations at the voice-coil, the mechanical part describes the force-velocity relations induced by the damped suspension of the driver diaphragm and its mass and finally the acoustic part describes the damped mass-spring-system behaviour of the vented box and the sound radiation from the diaphragm and vent. The corresponding diagrams for a loudspeaker in a closed box, in an infinite baffle or in open air (unbaffled) can be obtained by omission of the elements representing the vent and the box respectively [Thiele, 1971a, p.384]. A detailed description of the underlying theory of electro-acoustical networks and their application to loudspeaker models can be found in the famous book by Beranek [1959]. Further key papers were published by Thiele [1971a,b] and Small [1972, 1973a,b,c,d,e] who not only give an excellent literature review and summary on all relevant aspects of the topic but also derive methods for the measurement of the necessary loudspeaker parameters which allow a target-oriented design of the high-pass filter characteristics of a loudspeaker system. These parameters are widely known as the "Thiele-Small" parameters and are nowadays specified by most manufacturers of loudspeaker drivers. It is important to mention that the presented electrical analog network model is strictly speaking only applicable in the "piston range" [Thiele, 1971a] of a loudspeaker, which is defined as the frequency range below which the circumference of the speaker is less than the wavelength of the radiated sound:

$$f < \frac{c}{\sqrt{4\pi S_d}} \,. \tag{6.3}$$

This is due to the fact that for higher frequencies it is no longer suitable to model the mechanical and acoustical part of the loudspeaker by the lumped components given in figure 6.2. In particular the loudspeaker diaphragm and its suspension can no longer be approximated by an ideal piston with a damped spring suspension, due to the formation of eigenmodes on the loudspeaker diaphragm. Moreover, in the case of a boxed loudspeaker the formation of eigenmodes in the box also necessitates a more complex high frequency model than a simple compliance for the air cavity in the box.
Coming back to the field of room acoustic simulations we can conclude that the electrical analog network model of a loudspeaker system is a useful tool for the model of sound sources in room acoustic FE simulations, since it allows the calculation of the low frequency diaphragm velocity as a function of the "Thiele-Small" parameters, which are generally supplied by the loudspeaker manufacturer, and the dimensions of the considered loudspeaker box (so no additional measurements are necessary). On the other hand, in the case of room acoustic GA simulations, this model can only be used with considerable precaution, knowing that the prediction of the free field pressure transfer function for the mid- and high-frequency range contains non-negligible simplifications and that the model gives no indication about the directional characteristics of the loudspeaker system. Taking this into account we will confine ourselves to the low frequency application of the model in the FE domain.

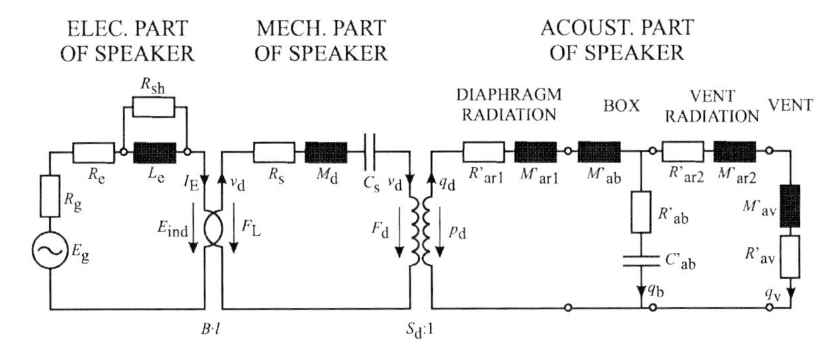

Figure 6.2.: Complete (electromechanical) acoustical circuit of loudspeaker in vented box (after Thiele [1971a]).
E_g = open-circuit voltage of audio amplifier; E_{ind} = voltage induced through voice coil movement; F_L = Lorenz force acting on voice coil; F_d = force acting on diaphragm; p_d = acoustic pressure acting on diaphragm; I_E = electric current through voice coil; v_d = velocity of diaphragm; q_d = volume velocity of diaphragm; q_b = volume velocity of box; q_v = volume velocity of vent or port; B = magnetic flux density in air gap; l = length of wire in air gap; S_d = equivalent piston area of diaphragm; R_g = amplifier output impedance; R_e = voice coil resistance; L_e = voice coil inductance; R_2 = electrical resistance due to eddy current losses; L_2 = para-inductance of voice coil; R_s = mechanical resistance of suspension; M_d = mechanical mass of diaphragm and voice coil; C_s = mechanical compliance of suspension; R'_{ar1} = acoustic radiation resistance for front side of loudspeaker diaphragm; M'_{ar1} = acoustic radiation mass (air load) for front side of loudspeaker; M'_{ab} = acoustic mass (air load) on rear side of loudspeaker; R'_{ab} = acoustic resistance of box; C'_{ab} = acoustic compliance of box; R'_{ar2} = acoustic radiation resistance of vent; M'_{ar2} = acoustic radiation mass (air load) of vent; M'_{av} = acoustic mass of air vent; R'_{av} = acoustic resistance of air in vent.

In this context an interesting question arises by asking whether the low frequency loudspeaker behaviour, namely its diaphragm velocity, is influenced in any way by the acoustic load that the sound field in the considered room exerts on the loudspeaker diaphragm. In the equivalent circuit diagram of figure 6.2 this influence is accounted for by the radiation impedance

$$\underline{Z}_{\text{rad}}(f) = \underline{Z}_{\text{ar1}}(f) \cdot S_{\text{d}} = (R'_{\text{ar1}}(f) + jM'_{\text{ar1}}(f)) \cdot S_{\text{d}}, \tag{6.4}$$

which is given by the ratio of the surface averaged complex sound pressure $\langle p_d \rangle$ and the normal velocity v_d at the loudspeaker diaphragm[2]. With regard to applications in small room acoustics the above raised question will be discussed in detail in section 7.3 on the basis of an FE simulation study for three exemplary, differently sized and acoustically relevant small spaces.

Although we will see from this study that at least for room acoustic applications the feedback of the room sound field on the diaphragm velocity appears to be negligible, the following paragraph gives, for the sake of completeness, an approach that allows the full

[2]A general definition of the radiation impedance is given by Mechel [1989, p.278] using the apparent power Π as:

$$\underline{\Pi} = \frac{1}{2}\underline{Z}_{\text{rad}} \oint_{S_{\text{d}}} |v_{\text{d}}|^2 \, \text{d}S \, . \tag{6.5}$$

Under the assumption of a surface constant diaphragm velocity v_d (amplitude and phase), which is reasonable for a loudspeaker in the "piston range", it can be shown that this definition is equivalent to $Z_{\text{rad}} = \frac{\langle p_d \rangle}{v_d}$, where $\langle p_d \rangle$ is the sound pressure averaged over the loudspeaker diaphragm.

coupling of the (electromechanical) acoustical circuit of a boxed loudspeaker and the room acoustic FEM simulation. Since the presented considerations can easily be extended to hydro- or structure-borne acoustics, where a coupling between sound source and sound field is much more important, possible useful applications might be found in these areas.

In a brute force approach the coupling between loudspeaker and room sound field can be modeled by a full inclusion of the equivalent network diagram of the loudspeaker in the room acoustic FEM simulation, which means that the excitation quantity in the FE simulation is no longer the diaphragm velocity but the input voltage at the voice-coil terminals. The formulation of the according FEM coupling conditions (which could not be found in any other publication) is given in the appendix of this thesis. However, in the case where only one loudspeaker is present in the room the coupling between loudspeaker and sound field can be considered in good approximation by a much simpler three step procedure as follows:

1. Run a finite element simulation for the considered room and loudspeaker position with unity velocity on the diaphragm of the loudspeaker. In this case the resulting average sound pressure on the diaphragm of the loudspeaker, corresponds directly to the radiation impedance of the loudspeaker, since $Z_{\mathrm{L}} = \frac{p_{\mathrm{diaphragm}}}{v_{\mathrm{diaphragm}}} = \frac{p}{1\,\mathrm{m/s}}$.
2. Use the so-obtained radiation impedance in the electrical analog network model of the loudspeaker system to determine the actual diaphragm velocity for 1 V input voltage at the voice-coil terminals.
3. Multiply the sound pressure in the room that was obtained from the FEM simulation at $v_{\mathrm{diaphragm}} = 1\,\mathrm{m/s}$ with the actual diaphragm velocity obtained in step 2, to obtain the sound pressure in the room for the considered loudspeaker driven at 1 V input voltage.

In comparison to the full inclusion of the loudspeaker circuit into the FEM calculation (according to the descriptions in the appendix) this procedure has the advantage, that it is possible to exchange the loudspeaker at a given position in the room without having to re-run the time-consuming FEM simulation.

Things look different if we consider several loudspeakers in the room, since in this case their can in principle be mutual coupling between the loudspeakers. This can be understood by considering that in the case of e.g. 2 subwoofers in a small room the diaphragm velocities of both independent driver units depend in a strict sense on the sound field in the room, which itself depends again on the diaphragm velocities of the driver units. This problem can be solved by using the full coupling approach presented in the appendix. However, also in this case the coupling only needs to be considered if the diaphragm velocity is affected by the change in radiation impedance compared to free field radiation, which, as we will see in section 7.3, is generally not the case for moderate sized rooms and “normal” loudspeakers.

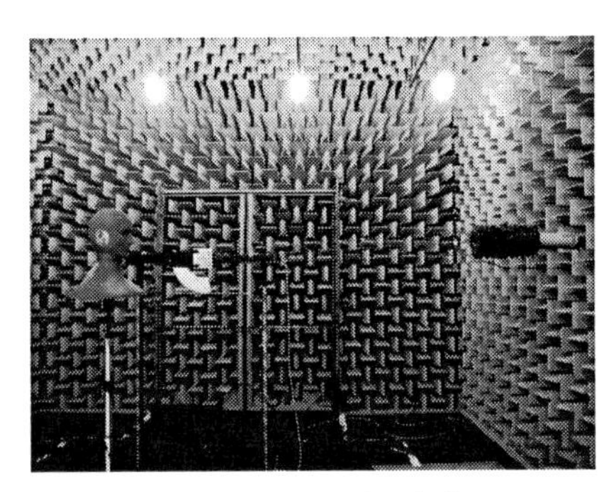

measurement upper hemisphere

measurement lower hemisphere

Figure 6.3.: Measurement setup for the measurement of the HRTFs of the ITA artificial head.

6.2. Binaural receiver characterization

In the course of this study measurements of the Head Related Transfer Functions (HRTFs) have been carried out for the ITA artificial head in the anechoic chamber at ITA of RWTH Aachen University. Figure 6.3 shows a picture of the measurement setup which consists of the ITA artificial head mounted at 2 m height on a turntable and a loudspeaker that moves on a swivel arm at a constant distance of 1.75 m around the head. The measurement setup is equivalent to the directivity measurement setup shown in figure 6.1 (c) except for the fact that for the HRTF measurements the artificial head is mounted on the turntable and a loudspeaker moves on the swivel arm. The floor-reflection in the measurements is canceled by a suitable time window. Possible low frequency errors (< 100 Hz) which are introduced by this time-window, can be corrected by hand by letting the HRTF data converge to 0 dB below 100 Hz, which is suitable under the reasonable assumption that the head does hardly influence the sound field at frequencies below 100 Hz. The measurements were carried out on a spherical Gaussian grid of order 70 which approximately corresponds to a two degree resolution. The Gaussian grid was chosen to allow the efficient calculation of a spherical harmonics representation of the HRTF data. This supplies an elegant way to interpolate the HRTF data at positions between the actual measurement points. The application of spherical harmonics framework to HRTF data and source directivity data is currently investigated in another study at ITA of RWTH Aachen University [Pollow et al., 2011]. However, to-date the simulation software *RAVEN* only supports the import of HRTF data on an equiangular grid. Since the positioning uncertainty of the measurement setup is estimated to be approximately $1°$, the HRTF data was interpolated with the nearest neighbor method to an equidistant $3°$ and $5°$ grid for the inclusion in the room acoustic GA software *RAVEN*.

The ITA artificial head uses a matched pair of free field equalized 1/2" Schoeps microphones which are placed at the blocked ear canal entrance of the left and right ear. In order to obtain the desired "free field transfer function" according to Blauert [1997] (cf.

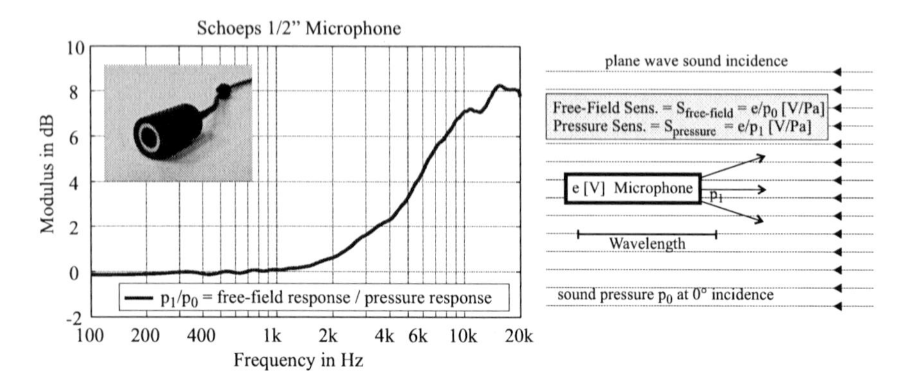

Figure 6.4.: Definitions of free field and pressure sensitivity of a microphone and the respective ratio for the used Schoeps microphones.

section 4.2.4) and also to cancel the frequency characteristics of the used loudspeaker and microphones the head measurements for the left and right ear are referenced to a measurement without the head, where the respective Schoeps microphone (used in the left and right ear) is placed at the centre position of the head pointing to the loudspeaker.

It is important to mention in this context that for the upper frequency range (> 4 kHz) diffraction effects at the diaphragm and the enclosure of the used 1/2" microphones cause a considerable modification to the sound field into which the microphones are inserted. Taking into account that in the head measurements the microphone housing is hidden in the artificial head while in the reference measurements it is fully exposed to the sound field, this implies that the undesired influence of the microphone housing does not cancel out in the ratio of the head and the reference measurements and thus needs to be corrected somehow in the post-processing of the HRTF measurements. In other words this means that in the case of the head measurement we want to measure the actual pressure at the blocked ear canal entrance, i.e. at the microphone diaphragm, but in the reference measurement we want to measure the pressure at the centre of the head when the microphone is not there. Such a correction can be effectuated by making use of the so-called microphone *free field correction*, which is given by the ratio of the *free field response* and *pressure response* of a microphone. "The *Free Field Response* of a microphone is the ratio of the RMS output voltage to the RMS sound pressure existing in a free field (plane sound waves) at the microphone location with the microphone removed. The *Pressure Response* of a microphone is the ratio of the RMS output voltage to the RMS sound pressure, uniformly applied over the diaphragm" (Taken from "Brüel & Kjaer, One-inch Condenser Microphones, 4131/32 - instructions and applications. Reprint July 1968"). At frequencies where the wavelength is smaller then ten times the diameter of the microphone, diffractions of the sound waves at the microphone start to cause considerable deviations between the free field and pressure response. For free field equalized microphones the diaphragm resonance is generally damped in a way that compensates the diffraction effects at the diaphragm for frontal incidence in order to achieve the flattest possible free field response. However, this leads to a considerable fall off in the pressure response at higher frequencies.

Figure 6.4 illustrates the definitions of the *free field* and *pressure response* of a microphone and shows the respective ratio for the used Schoeps microphones. An elegant way to measure the free field correction for a given microphone is described in the appendix of this thesis. This method was also used to obtain the curve for the Schoeps microphones. However, it should be mentioned that for many measurement microphones this curve can be obtained directly from the calibration charts supplied by the manufacturer. Summing up, the HRTF for a given angle of incidence can be calculated as

$$\mathrm{HRTF}(\theta,\varphi) = \frac{p_{\text{head}}}{p_{\text{reference}}} = \frac{e_{\text{head}}}{e_{\text{reference}}} \cdot \frac{S_{\text{free field}}}{S_{\text{pressure}}}, \tag{6.6}$$

where p_{head}, $p_{\text{reference}}$, e_{head}, $e_{\text{reference}}$ are the RMS values of the pressure and microphone output voltages for the head and the reference measurement respectively and S_{pressure}, $S_{\text{free field}}$ are the pressure sensitivity and the free field sensitivity (for normal sound incidence) of the microphone. An alternative approach to avoid this problem would for example consist of using high quality miniature microphones that have an almost equal free field and pressure sensitivity for the whole audible frequency range. The used microphones should however have an acceptably low self-noise level.

7. Simulation studies on selected aspects

The present section discusses selected questions that have been raised in the previous chapters but could not be easily resolved on the basis of the theoretical considerations given in these sections. These questions relate to the use of the image source method for low frequency prediction in concave rooms, the efficient modeling of porous absorbers in room acoustic FE simulations and the coupling effects between low frequency sound fields and the loudspeaker diaphragm velocity. Our results are based on comparative simulation studies, which in the case of the porous absorber study are backed with measurement results obtained in the respective room.

The first study presented in section 7.1 discusses the interesting question brought up in section 4.2.1 regarding the applicability of the image source method to the prediction of the modally dominated part of an RTF in concave rooms with arbitrary boundary conditions on the room walls. The study is based on the comparison of low frequency RTFs obtained from the image source and the finite element method for a simple room geometry with complex boundary conditions, where the FE results are considered as the reference. In order to allow the image source calculation up to very high orders ($n_{\text{IS}} > 50$) the study uses an efficient image source modeling algorithm for cuboid rooms which allows the assignment of individual frequency and angle dependent complex reflection factors to each of the 6 walls in the room.

The second simulation study presented in section 7.2 focusses on the comparison of room acoustic FE simulations and measurements in a simple, hexahedral scale model room. Due to the simplicity of the simulation and measurement setup (room geometry, boundary materials and source characteristics) this well-controlled scale model environment allows a measurement backed investigation of isolated aspects regarding the FE simulation quality. In particular, simulations and measurements are conducted in the empty scale model room and in the room with porous absorber panels applied to three of the six room walls. While the former constitutes a kind of "best-case scenario", which shows the potential of the FE method to realistically predict the low frequency sound field in a room, the latter focusses on the efficient modeling of the absorber panels in the room acoustic FE simulations and their influence on the overall simulation result.

Finally, section 7.3 discusses the question raised in section 6.1.2 regarding the possible coupling between the room sound field and the loudspeaker behaviour, namely its diaphragm velocity. This topic is investigated based on an FE simulation study using an exemplary low- to midrange loudspeaker unit, which is described by its Thiele-Small parameterization and three differently sized, acoustically relevant spaces. Special emphasis is put on the order of magnitude of the low frequency radiation impedance of a loudspeaker that radiates into a small room compared to the other components in the acoustical loudspeaker circuit and the dependence of the radiation impedance on the size, shape and damping of the considered room.

7.1. Using the image source method for low frequency prediction

The present section investigates the applicability of the image source method to the prediction of low frequency sound fields in small rooms. The fundamentals, potentials and limitations of the method were already discussed at length in section 4.2.1 and shall not be repeated here. The simulation study focusses on three major aspects: (a) The quantification of the errors that are introduced by the mirror source approximation in the case of non-ideally rigid or soft walls (see section 3.2.3) as a function of frequency and average absorption in the room; (b) the simulation error due to the truncation of the IS calculation based on different interrupt criteria; (c) The impact of commonly applied modifications to the ISM method (e.g. negligence of angle dependance of the reflection factor, negligence of the phase of the reflection factor and the time shift of single mirror source contributions to full sample positions). In order to investigate these aspects ISM simulations are compared with FEM simulations in a simple cuboid room for different boundary conditions. All FEM simulations for this study were run with *LMS Virtual Lab Rev9-SL3.*

Since the calculation of high order image sources in arbitrarily shaped rooms is a very computation intensive task and the current version (September, 2011) of the simulation software *RAVEN* can even for the simplest rooms only calculate ISs up to the 8th reflection order, an efficient image source modeler was implemented for the special case of cuboid rooms.

7.1.1. Image source modeler for a cuboid room

For the present study an optimized *CAFR* (Complex Angle and Frequency dependent Reflection factor) image source method was implemented for the special case of cuboid rooms, which allows the assignment of individual reflection factors $\underline{R}(\theta)$ to each of the six room walls. The impulse response generation consists of three major steps, which are summarized in the following:

1. **Determination of image source positions**

   In a cuboid room the image sources are arranged on a simple regular pattern, which allows the efficient calculation of image sources up to very high orders ($n > 50$). Moreover a visibility test is not necessary in the case of a cuboid room, since each image source position on the regular grid corresponds to exactly one visible image source. If we assume a cuboid room with dimensions (L_x, L_y, L_z) which has its centre at $P_0 = [0, 0, 0]$ and the source and receiver positions at $P_\text{s} = [x_\text{s}, y_\text{s}, z_\text{s}]$ and $P_\text{r} = [x_\text{r}, y_\text{r}, z_\text{r}]$ respectively (see figure 7.1) than the position of an image source of order (n_x, n_y, n_z) and its distance to the receiver are given by

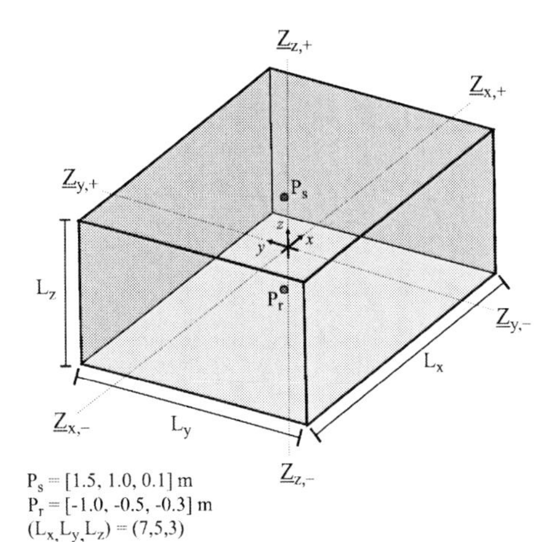

Figure 7.1.: Dimensions, coordinate and variable definition for the cuboid room used for the IS simulation study.

$$d_{\mathrm{IS,r}}(n_x, n_y, n_z) = \| P_{\mathrm{r}} - P_{\mathrm{IS}}(n_x, n_y, n_z) \|, \quad \text{with} \tag{7.1}$$

$$P_{\mathrm{IS}}(n_x, n_y, n_z) = \begin{bmatrix} x_{\mathrm{IS}} \\ y_{\mathrm{IS}} \\ z_{\mathrm{IS}} \end{bmatrix} = \begin{bmatrix} n_x L_x + ((-1)^{n_x} \cdot x_{\mathrm{s}}) \\ n_y L_y + ((-1)^{n_y} \cdot y_{\mathrm{s}}) \\ n_z L_z + ((-1)^{n_z} \cdot z_{\mathrm{s}}) \end{bmatrix}, \tag{7.2}$$

where the order indices $n_{x/y/z}$ are defined as indicated in figure 7.1.

2. **Determination of attenuation of each individual image source**

In order to calculate the contribution of a single image source to the overall room impulse responses we have to consider that each wall reflection can be modeled by a frequency domain multiplication of the excitation spectrum with the complex and angle of incidence dependent reflection factor $\underline{R}(\theta)$. In addition to the sound attenuation at the room boundaries the image source contribution also has to be attenuated by $\frac{1}{d_{\mathrm{IS,r}}}$ in order to account for the sound decay of a spherical pressure wave propagating in free space. In mathematical notation the attenuation factor A_{IS} of an image source of order (n_x, n_y, n_z) can thus be given as

$$\begin{aligned} A_{\mathrm{IS}}(n_x, n_y, n_z) &= \tfrac{1}{d_{\mathrm{IS,r}}} \cdot \ldots \\ (\underline{R}_{\mathrm{x},+} \cdot \underline{R}_{\mathrm{x},-})^{\left\lfloor \frac{|n_x|}{2} \right\rfloor} &\cdot \left(\tfrac{1+\mathrm{sign}(n_x)}{2} \underline{R}_{\mathrm{x},+} + \tfrac{1-\mathrm{sign}(n_x)}{2} \underline{R}_{\mathrm{x},-} \right)^{(n_x \bmod 2)} \cdot \ldots \\ (\underline{R}_{\mathrm{y},+} \cdot \underline{R}_{\mathrm{y},-})^{\left\lfloor \frac{|n_y|}{2} \right\rfloor} &\cdot \left(\tfrac{1+\mathrm{sign}(n_y)}{2} \underline{R}_{\mathrm{y},+} + \tfrac{1-\mathrm{sign}(n_y)}{2} \underline{R}_{\mathrm{y},-} \right)^{(n_y \bmod 2)} \cdot \ldots \\ (\underline{R}_{\mathrm{z},+} \cdot \underline{R}_{\mathrm{z},-})^{\left\lfloor \frac{|n_z|}{2} \right\rfloor} &\cdot \left(\tfrac{1+\mathrm{sign}(n_z)}{2} \underline{R}_{\mathrm{z},+} + \tfrac{1-\mathrm{sign}(n_z)}{2} \underline{R}_{\mathrm{z},-} \right)^{(n_z \bmod 2)}, \end{aligned} \tag{7.3}$$

where 'mod' is the modulo operator and the complex reflection factors $\underline{R}_{\mathrm{x/y/z},+/-}$, which are assigned to the six room walls according to figure 7.1, are of course angle dependent. The corresponding angles of incidence $\theta_{\mathrm{x/y/z}}$ are given by

$$\theta_x(n_x, n_y, n_z) = \arccos\left(\frac{|x_{\text{IS}} - x_{\text{r}}|}{d_{\text{IS,r}}}\right) \tag{7.4}$$

$$\theta_y(n_x, n_y, n_z) = \arccos\left(\frac{|y_{\text{IS}} - y_{\text{r}}|}{d_{\text{IS,r}}}\right) \tag{7.5}$$

$$\theta_z(n_x, n_y, n_z) = \arccos\left(\frac{|z_{\text{IS}} - z_{\text{r}}|}{d_{\text{IS,r}}}\right), \tag{7.6}$$

where θ_x, θ_y and θ_z are the angles of incidence on the walls perpendicular to the x-, y- and z-axis respectively. It is important to mention that in a rectangular room for each individual image source only three different angles of incidence exist. This means that for a given image source at $P_{\text{IS}}(n_x, n_y, n_z)$ and receiver point P_{r} all reflections at the walls perpendicular to the x-axis have the same angle of incidence $\theta_x(n_x, n_y, n_z)$. The same holds for the reflections at the walls perpendicular to the y- and z-axis.

3. **Calculation of each image source contribution to the RIR**

   With the consideration of frequency dependent reflection factors each image source contributes a delayed, attenuated and widened impulse to the overall impulse response. Using the notations from above this impulse response can be given by the summation of all individual image source contributions as follows:

$$\underline{p}_{\text{r}}(n_x, n_y, n_z) = \sum_{-n_{\text{x,max}}}^{n_{\text{x,max}}} \sum_{-n_{\text{y,max}}}^{n_{\text{y,max}}} \sum_{-n_{\text{z,max}}}^{n_{\text{z,max}}} \underline{\hat{p}}\, A_{\text{IS}}(n_x, n_y, n_z)\, e^{\frac{-j\omega d_{\text{IS,r}}(n_x, n_y, n_z)}{c}}. \tag{7.7}$$

   The above given frequency domain formulation has the positive side effect that the distance related part of the time delay of each individual image source impulse is specified with sub-sample precision by the term $\exp(\frac{-j\omega d_{\text{IS,r}}}{c})$, which in the time domain corresponds to a sampled sinc-function as follows:

$$\underline{p}_{\text{IS}} = \underline{\hat{p}}\, A_{\text{IS}} \operatorname{sinc}\left(\left(t - \frac{d_{\text{IS,r}}}{c}\right) \cdot f_{\text{sampling}}\right). \tag{7.8}$$

For the sake of completeness it should be mentioned that in the case of complex frequency dependent wall impedances equation 7.8 becomes more complicated since the sinc-function needs to be convoluted with the attenuation spectrum A_{IS}.

In order to verify the implemented image source algorithm our results were compared to the freely available image source algorithm of Lehmann and Johansson [2008][1], which uses strictly negative real valued reflection coefficients in order to enforce a zero DC offset in the impulse response (since this method results in a negative sign of all image source impulses with uneven order in the RIR). Since this method considerably changes the low frequency RTF, by applying a 180° phase shift at every boundary reflection, the code by Lehmann was changed to also allow for positive real-valued reflection coefficients. The so obtained results from our simulations and those run with the modified Lehmann code for different real-valued, positive, frequency independent reflection coefficients on the room

[1]The *MATLAB* based IS implementation by Lehmann and Johansson [2008] can be downloaded under the URL `http://www.eric-lehmann.com/`.

walls showed a perfect agreement. However, it is emphasized that the present implementation extends the Lehmann implementation by consideration of complex, frequency and angle dependent reflection factors. A verification of these extended features is given in the following section.

7.1.2. Verification of algorithm for complex valued, frequency and angle dependent reflection factors

In a first step an IS and FE simulation were run for a cuboid room with different complex and frequency dependent surface impedances $\underline{Z}_S$ assigned to the 6 boundary walls. In the simulations the room, source and receiver setup of figure 7.1 was used. The comparison of the results serves two main purposes: On the one hand the implemented IS algorithm shall also be verified in the case of complex valued, frequency and angle dependent reflection factors and on the other hand a first assessment of the potential of the ISM to predict the modally dominated part of a room transfer function shall be obtained.

In the FE simulation we therefore use the surface impedances as boundary conditions while in the IS simulation we use equation 3.9 to calculate the angle dependent reflection factors for each image source. Figure 7.2 shows the normal incidence reflection factors which were calculated from the surface impedances assigned to the room boundaries. In the case of the two frequency dependent boundary conditions the two-port network model for layered absorbers described in section 5.3 was used to calculate the surface impedances of the heavily damped Helmholtz resonator and the single-layer porous absorber. The corresponding parameterization of the absorber layers is also given in figure 7.2.

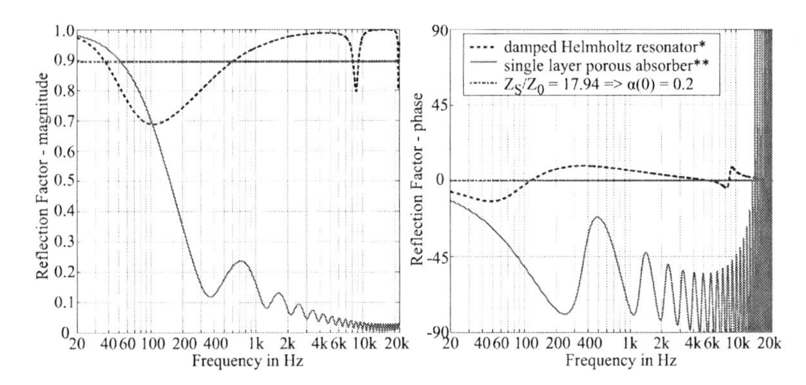

Figure 7.2.: Normal incidence reflection factors $R(0)$ calculated from the surface impedances Z_S that were used in the simulations shown in figure 7.3.

The complex surface impedances for the single layer porous absorber and the damped Helmholtz resonator were calculated using the two-port network model for layered absorbers described in section 5.3, with the following parameterization.

* Damped Helmholtz resonator (3 layers): (1) perforated panel with $t = 10\,\mathrm{mm}$, $\sigma_s = 0.0314$, $d = 20\,\mathrm{mm}$; (2) mass with flow resistivity with $t = 1\,\mathrm{mm}$, $\rho_m = 80\,\mathrm{kg/m^3}$, $\Xi = 20\,\mathrm{kPas/m^2}$; (3) porous absorber (Komatsu model) with $t = 100\,\mathrm{mm}$, $\Xi = 10\,\mathrm{kPas/m^2}$

** Single layer absorber (1 layer): (1) porous absorber (Komatsu model) with $t = 200\,\mathrm{mm}$, $\Xi = 5\,\mathrm{kPas/m^2}$

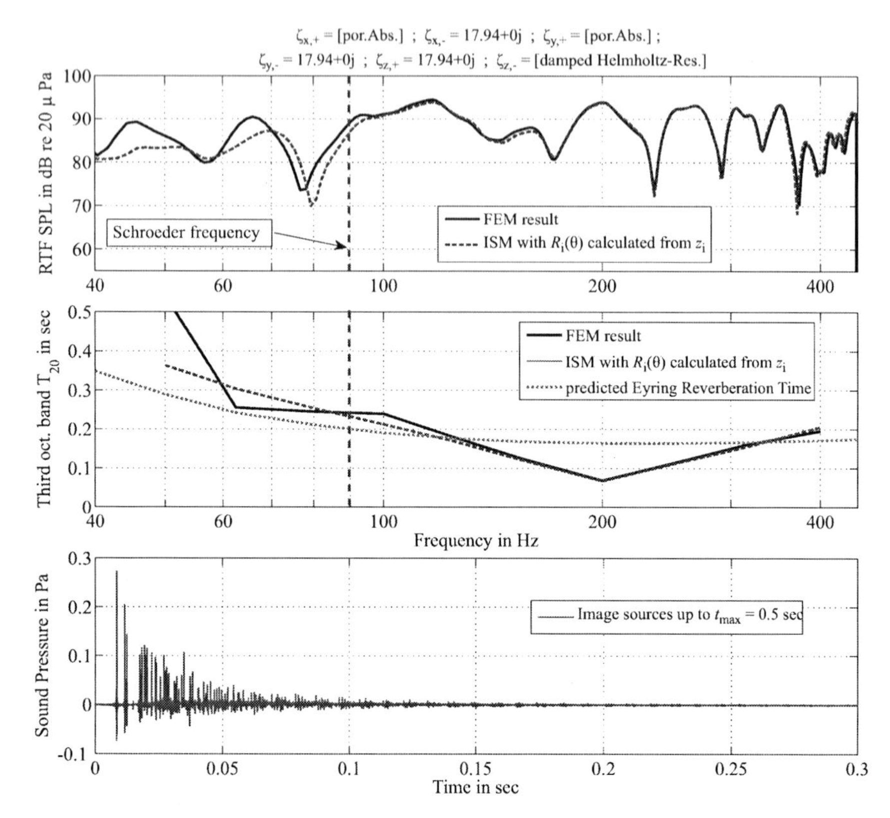

Figure 7.3.: Comparison of low frequency room transfer function and reverberation time in a cuboid room obtained with the FE and IS method for distributed, frequency dependent, complex impedances on the room walls.
The bottom plot shows the broadband IR obtained from the IS method. The IS method uses a frequency and angle dependent reflection factor, which is calculated from the frequency dependent, complex impedances on the six room walls. The reflection factor of the used porous absorber and the damped Helmholtz resonator are given in figure 7.2. The normalized impedance on the other three walls ($\zeta_i = 17.94$) corresponds to an absorption of 0.2 for normal sound incidence.

In order to avoid any artefacts from a too early truncation of the impulse response the truncation time for the image source order has been set to a very conservative value of $t_{\text{trunc}} = 0.5\,\text{s}$, which exceeds the predicted Eyring reverberation time for the whole audible frequency range. Additionally it has been verified that a further increase of t_{trunc} does not noticeably change the obtained ISM room transfer function. It can thus be stated that the remaining differences between the FEM and ISM calculation are only caused by the inherent errors in the image source method itself. Figure 7.3 shows the results of the simulations. It can be seen that in the considered case an excellent match between the transfer functions and reverberation times from both methods is obtained for frequencies above the Schroeder frequency f_S. Below f_S the FEM and the ISM results slowly drift apart. However, the dominant dips and peaks caused by the fundamental room modes are still roughly covered by the ISM prediction even at these very low frequencies.

This interesting result agrees well with the error plots for a single image source reflection shown in figure 3.3, which indicate that the error increases significantly if the sum of heights of source and receiver above the reflecting plane becomes less than a couple of wavelengths. Thus if the sound source or the receiver are close to any room wall in terms of wavelengths (or in other words 'at sufficiently low frequencies') the image source approximation will lead to errors in the first order reflections which also propagate to the higher reflection orders. The dependence of this error on the average absorption in the room will be discussed in the next section.

7.1.3. Error due to truncation at maximum IS order based on different interrupt criteria

The results of the previous section show that if complex surface impedance data is available for all room boundaries and the image sources can be determined up to sufficiently high orders, the *CAFR-ISM* is capable of predicting the modal structure of the low frequency sound field in concave rooms with high accuracy, except for the very low end of the frequency range. Taking into account the considerable computational complexity of high order image source determination in arbitrary rooms, it is now interesting from a practitioners point of view to investigate, up to which image source order the results need to be calculated at minimum to get a stable low frequency response and how suitable interrupt criteria can be defined for the IS method.

In order to approach this question three differently damped rooms ($\alpha_{\mathrm{diff},1} = 0.21$, $\alpha_{\mathrm{diff},2} = 0.41$, $\alpha_{\mathrm{diff},3} = 0.71$) and two different interrupt criteria are used for the image source calculation. With regard to the room damping, the exemplary values of the surface impedances are again taken from the paper by Suh and Nelson [1999] in order to allow comparison between the results in section 3.2.3. With regard to the interrupt criteria the image sources are calculated up to (a) different fractions of the predicted Eyring reverberation time T_E and (b) different maximum attenuation factors relative to the direct sound. In particular, the attenuation interrupt criterium discards all image sources with $10\log_{10}(\frac{|e_{\mathrm{direct}}|}{|e_{\mathrm{IS}}|}) > A$, where e_{direct} and e_{IS} are the energies of the direct sound impulse and the considered image source reflection impulse and A is the attenuation threshold in dB. The impulse energies are calculated as the squared sum over all time samples unequal zero. In the case of frequency independent absorption, where the image source contributions are dirac impulses, the energies are simply given by the squared dirac impulse amplitude. The results for both interrupt criteria are given in figures 7.4 and 7.5 respectively. It can be seen that irrespective of the average room absorption the low frequency room impulse response becomes stable at about $t_{\mathrm{max}} = 0.3\,T_E$ in the case of the first and for $A_{\mathrm{max}} = 40 - 50\,\mathrm{dB}$ in the case of the second interrupt criterium. The fundamental modal structure is however already well captured at even smaller values of these interrupt criteria ($t_{\mathrm{max}} = 0.2\,T_E$ and $A_{\mathrm{max}} = 30 - 40\,\mathrm{dB}$). The highest reflection orders and the total number of image sources for these values of the interrupt criteria obviously depend on the average absorption in the room and are also given in figures 7.4 and 7.5.

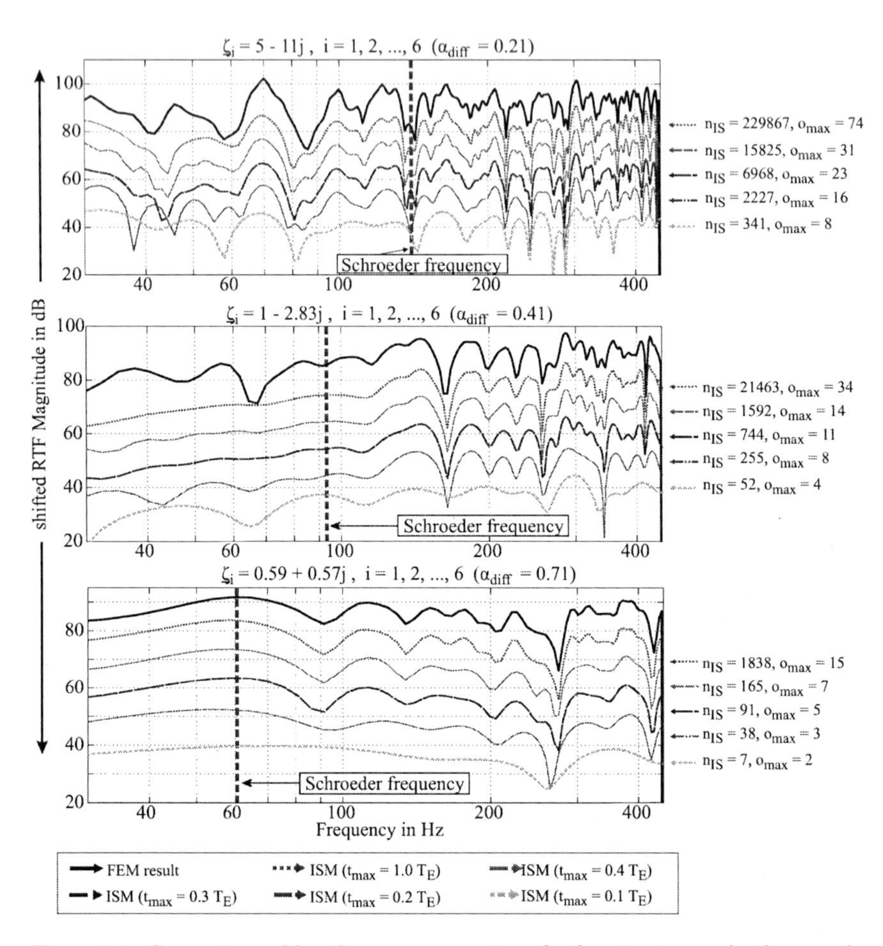

Figure 7.4.: Comparison of low frequency room transfer function in a cuboid room obtained with FE and IS method for three different average absorption values (RT interrupt criterium).

The IS method is calculated up to different truncation times, which correspond to fractions of the predicted Eyring reverberation time. The highest image source order $o_{\max}$ for each truncation time as well as the total number of considered image sources n_{IS} is given in the legend. The RTF magnitudes are shifted by 10 dB for better distinction of the curves.

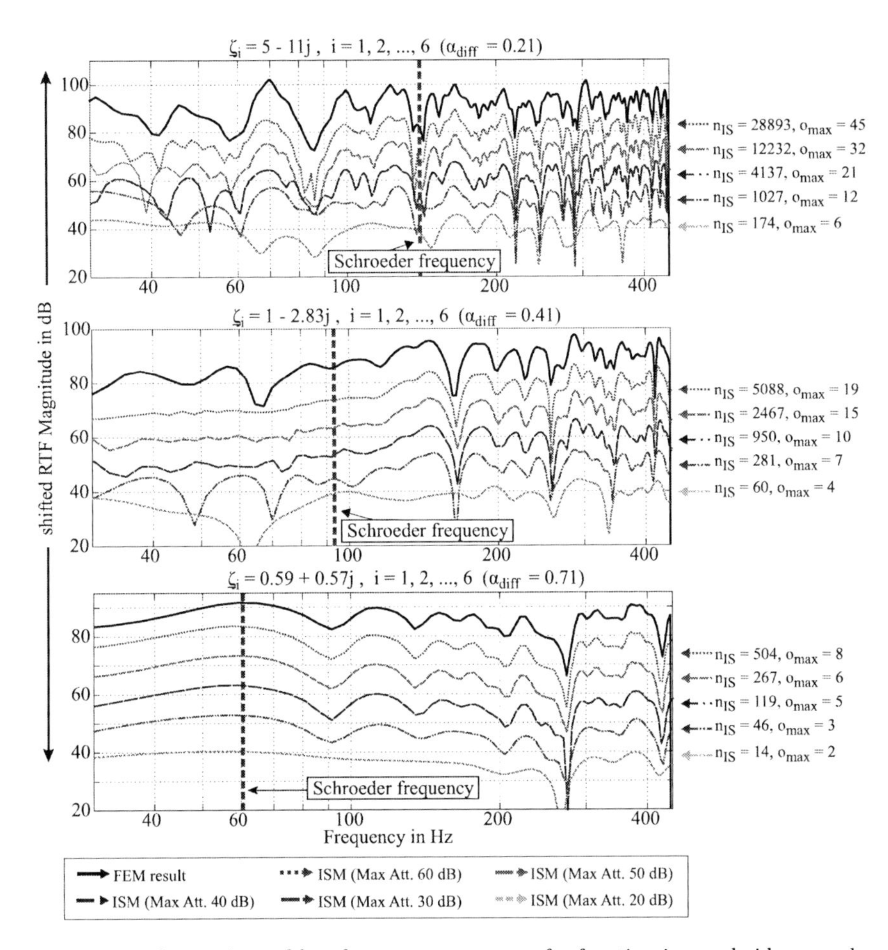

Figure 7.5.: Comparison of low frequency room transfer function in a cuboid room obtained with FE and IS method for three different average absorption values (Attenuation interrupt criterium).

The IS method only considers image sources reflections whose magnitude exceed a given threshold level which is given relative to the direct sound level. The IS curves are plotted for 5 different thresholds and the highest order image source o_{max} for each threshold as well as the total number of considered image sources n_{IS} is given in the legend. The RTF magnitudes are shifted by 10 dB for better distinction of the curves.

By looking at the reference ISM results with $t_{\mathsf{max}} = 1.0\,T_E$ another interesting observation can be made. Firstly, it can be seen that irrespective of the average room absorption the ISM shows a very good match with the FEM results above the Schroeder frequency f_S. However, below f_S the highest deviation is found for an average absorption of $\alpha_{\mathsf{diff}} = 0.41$. Although it was shown in section 3.2.3 that the error of the image source approximation increases with increasing wall absorption and thus the biggest error of the image source approximation is expected at $\alpha_{\mathsf{diff}} = 0.71$, this observation can be explained by taking into account that at very high average absorption levels the direct sound (which is modeled without error) becomes more and more dominant in the RIR.

Summing up, it can be concluded that the ISM already gives a good representation of the low frequency sound field at reasonable image source orders except for very low frequencies (inherent IS error) and for very low average room absorption (high orders needed). The *CAFR* ISM therefore offers an interesting possibility to fill the gap between the validity ranges of room acoustic FE and standard hybrid GA tools. Although the present results are obtained for a cuboid room with only one material property per room wall, it is believed that the above made reasoning is generally also valid for more complex concave rooms with oblique angled walls. However, it has to be kept in mind that the present results give no indication on the influence of convex edges in the room, which will introduce further errors into the ISM calculations. Moreover, considerable diffraction effects may also appear at 180° joints between different neighboring materials on a flat wall (e.g. an absorber mat which is mounted flush into a hard reflective panel). The influence of these diffraction effects on the modal structure of the sound field therefore needs to be further investigated in a future study.

7.1.4. Impact of commonly applied modifications to the ISM method

State-of-the-art hybrid geometrical acoustics tools often use modified versions of the above presented *CAFR* image source method. These modifications generally concern the used sound field descriptor (complex sound pressure vs. pressure magnitude or energetic quantities), the definition of the acoustic boundary conditions (absorption coefficient vs. complex reflection factor; angle dependent vs. diffuse field averaged quantities) or the way the individual image source contributions are inserted in the room impulse response (reflection delay in sample or sub-sample resolution; sign of image source impulses). The present section investigates the impact of these commonly found modifications on the modal prediction quality of the ISM. In particular, the following aspects are investigated:

1. The use of angle dependant reflection factors vs. the use of incidence-angle averaged reflection factors assuming field incidence.
2. The impact of the negligence of the reflection factor phase.
3. The effect of time-shifting the image source impulses to full sample positions.

Two simulation sets were run using again the room setup shown in figure 7.1 with a homogeneous and frequency constant surface impedance assigned to all 6 room walls in the FEM simulations. For the first simulation set we use $Z_{\mathrm{S,I}} = 17.94\,Z_0$ and for the second set $\underline{Z}_{\mathrm{S,II}} = (1 + 4\mathrm{j})\,Z_0$, respectively. Both impedances yield an absorption coefficient for normal incidence of $\alpha(0) = 0.2$. Thus the first impedance can be calculated from the second by setting the phase of the reflection factor $\underline{R}_{\mathrm{II}}(0)$ to zero.

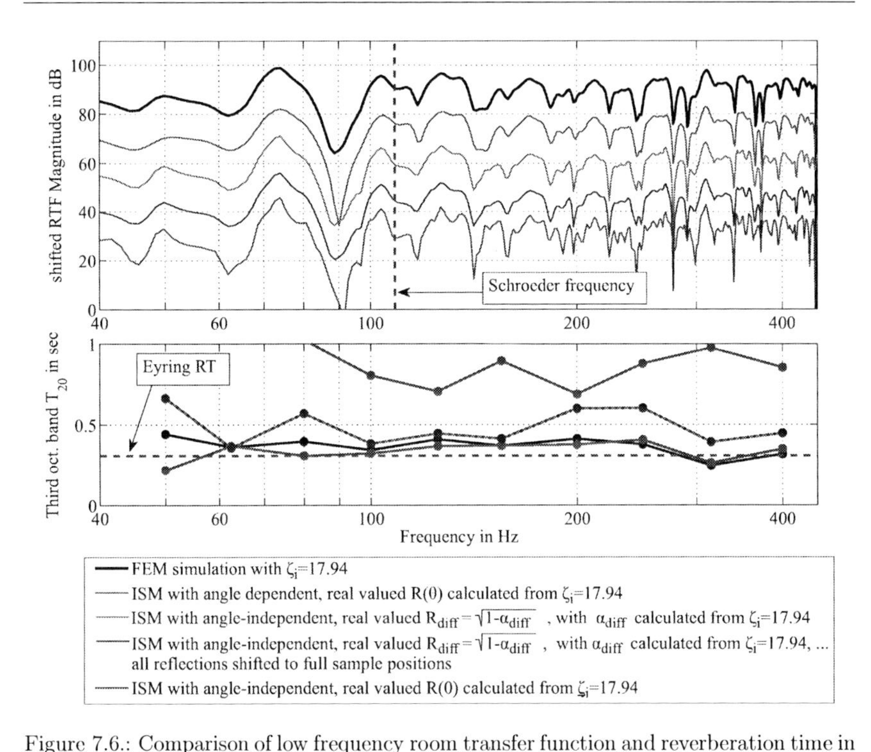

Figure 7.6.: Comparison of low frequency room transfer function and reverberation time in a cuboid room obtained with FE and IS method for a homogeneous, frequency constant and real impedance on all room walls.
The IS method uses different boundary representations (see legend). The magnitude of the differently obtained RTFs are shifted by 15 dB for better distinction of the curves.

Figure 7.6 shows the results obtained for the real-valued impedance $Z_{\mathrm{S,I}} = 17.94\,Z_0$. It can be seen, that except for the simulation with angle-independent reflection factor $\underline{R}(0)$, which shows a considerable overestimation of the reverberation time in the room, all other ISM simulations generally agree well with the FEM results. The best match is found as expected for the full *CAFR* method, but the simulations with the angle-independent diffuse field reflection coefficient $R_{\mathrm{diff}} = \sqrt{1-\alpha_{\mathrm{diff}}}$ also predict the modal characteristics of the room very well, although the reverberation time is slightly overestimated for frequencies above 160 Hz. Moreover it can be concluded from the figure that a time shift of the image source reflections to full sample positions does not deteriorate the ISM result noticeably. This result can be used for performance considerations, since it allows to calculate each image source sound impulse at a much lower FFT degree than the full room impulse response and than simply sort the snippets for each reflection into the full IR at the right sample position.

The results using the complex surface impedance $\underline{Z}_{\mathrm{S,II}} = (1+4\mathrm{j})\,Z_0$ in the FEM simulations are shown in figure 7.7 and again a very good agreement between FEM and the *CAFR* ISM results is found. In addition to the full *CAFR* method an ISM simulation was

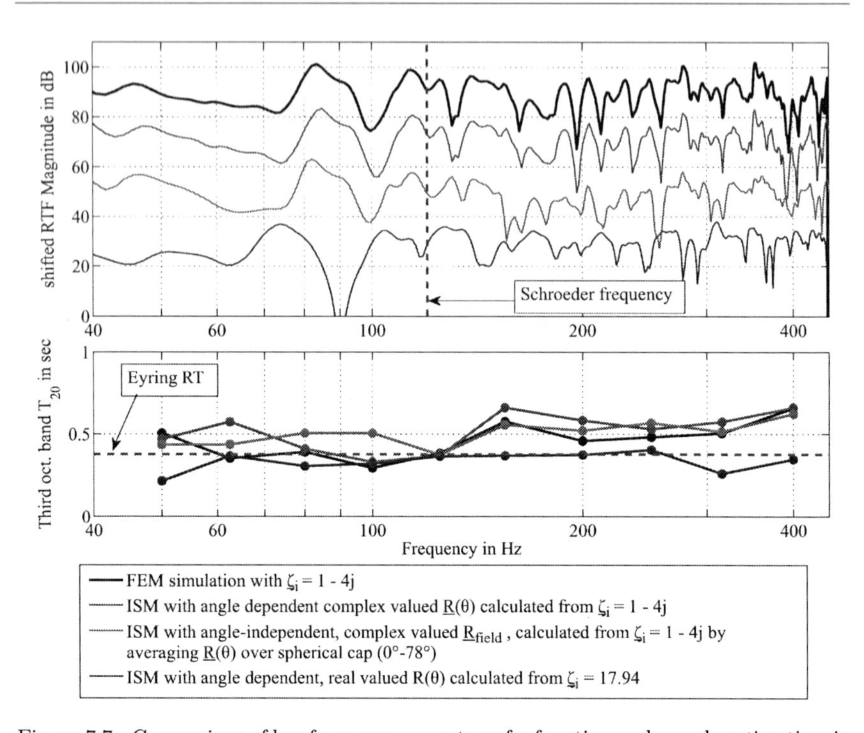

Figure 7.7.: Comparison of low frequency room transfer function and reverberation time in a cuboid room obtained with FE and IS method for a homogeneous, frequency constant and complex impedance on all room walls.
The IS method uses different boundary representations. The magnitude of the differently obtained RTFs are shifted by 20 dB for better distinction of the curves.

run that uses an angle-averaged complex reflection factor, which shall be denoted 'field incidence reflection factor' $\underline{R}_{\text{field}}$ and which is calculated by averaging the angle dependent values of $\underline{R}(\theta)$ over a spherical cap from $\theta = 0^\circ - 78^\circ$[2]. It can be seen that the results using this field incidence reflection factor roughly reflect the modal structure of the FEM simulation, but the results do not reach the quality of the *CAFR* results.

[2]The field incidence reflection coefficient is calculated as follows:

$$\underline{R}_{\text{field}} = \frac{\int_{0^\circ}^{78^\circ} \underline{R}(\theta)\sin(\theta)\,\mathrm{d}\theta}{\int_{0^\circ}^{78^\circ} \sin(\theta)\,\mathrm{d}\theta} \quad , \tag{7.9}$$

Since ideally diffuse conditions can generally not be obtained in typical rooms, the term 'field incidence' has emerged and is often referred to in building acoustics (cf. for example [Hopkins, 2007, p. 426ff]). Field incidence assumes that angles of incidence are restricted to being equally probable between 0°-78°. Hopkins [2007] states that the upper boundary for the angles of incidence is mostly empirically motivated and somehow lacks a physical meaning. However, the value of 78° has been adopted in many publications because of its commonly observed good fit with measured data.

Finally, we have also included the results using the *CAFR* method for the real valued impedance $Z_{S,I} = 17.94\,Z_0$ in figure 7.7. This result corresponds to a negligence of the reflection factor phase in the ISM simulation, which is one of the most common simplifications of the ISM method, since in most cases only absorption coefficients are known for the boundary materials and the reflection factor is calculated by $R(\theta) = \sqrt{1 - \alpha(\theta)}$. It can be seen that by setting the reflection factor phase to zero the modal structure of the low frequency sound field is considerably changed.

It can thus be concluded that the prediction of the modal structure of the sound field using the ISM is generally only possible if complex valued surface impedances are used as boundary input data. Only in the case of rather hard reflecting walls the use of real valued boundary data is admissible, since in this case the reflection factor phase is close to zero anyway. Additionally it can be stated that it is always preferable to account for the angle dependence of the reflection factor instead of using diffuse field averaged values, since if the complex surface impedances are known and the boundaries are considered locally reacting (with $\underline{Z}_S \neq f(\theta)$) the inclusion of the angle dependency comes at a low additional computational cost.

7.2. Room acoustic FE simulations in a scale model environment[3]

In order to investigate the potential of the FE method to realistically predict the low frequency sound field in small rooms, the present study compares measured and simulated room transfer functions in a well-controlled scale model environment with a hexahedron shape and variable boundary conditions. Simulations and measurements are carried out both for 'empty' (all walls hard and reflective) and 'damped' (3 walls lined with porous absorber panels) conditions. While the results for the empty condition mainly serve to validate the used FE simulation setup and source model, the investigations in the damped room specifically focus on the efficient modeling of porous absorbers in room acoustic FE simulations. The acoustic characteristics of a porous absorber panel are therefore modeled by using both impedance boundary conditions for normal and field incidence as well as a coupled 3D-FE absorber model. Taking into account that using a coupled 3D absorber model is much more costly in terms of computation time and model complexity compared to an impedance boundary condition, it is worthwhile investigating in which cases an actual benefit can be expected from using such complex models.

Before presenting the measured and simulated results for the scale model room, the following section 7.2.1 gives a short literature review on the modeling of sound propagation in porous absorbers and section 7.2.2 discusses the models which have been used in the course of this study. Sections 7.2.3 - 7.2.4 then explain how the chosen models can be applied in room acoustic FE simulations (either as an impedance boundary condition or as a 3D-FE absorber model) and section 7.2.5 describes the according measurement and simulation setups. Finally, section 7.2.6 presents the simulation and measurement results obtained in the scale model room for the different conditions and section 7.2.7 concludes the study.

7.2.1. Review of porous absorber models and their applicability in room acoustic FE simulations

In room acoustic FE applications porous absorbers can either be modeled by an appropriate impedance boundary condition or by a full 3D-FE absorber model, which is coupled to the FE fluid domain. As was already discussed in section 5 acoustic surface impedances

[3]Similar investigations on comparative measurements and simulations in the considered scale model room were already presented in Aretz [2009] (empty scale model room) and Aretz and Vorländer [2010] (damped scale model room) and the descriptions of the used measurement and simulation setups as well as the theoretical part considering the used absorber models have been taken mostly from these earlier publications. Please note that for better readability no quotation marks are used to mark sentences or paragraphs which are identical to the corresponding parts in Aretz and Vorländer [2010]. It is emphasized however, that the present elaboration on the topic has been enriched with a complete new set of measurement results in the scale model room as well as new simulation results with refined boundary conditions for the hard and reflective MDF walls. Altogether, the extended results presented in this section corroborate the fundamental conclusions given in the earlier publications and provide some further new insights, especially with regard to the modeling of the hard reflective MDF walls.

of porous absorbers can either be determined by measurement or by calculation based on a physical model of the wave propagation inside the absorber. These physical models can be roughly divided into models considering a rigid or an elastic absorber skeleton. A well-established model for porous absorbers with a rigid skeleton is the so-called 'equivalent homogeneous fluid' (EHF) approach, which owes much of its derivation to the theory of lossy electric transmission lines. This theory models a porous absorber as a damped fluid medium, where the complex propagation constant $\underline{\Gamma}_{\mathrm{a}}$ and the characteristic impedance $\underline{Z}_{\mathrm{a}}$ are derived as a function of either physically, phenomenologically or empirically based absorber parameters.

The original theory and the corresponding relations for $\underline{\Gamma}_a$ and $\underline{Z}_a$ were introduced by Zwikker and Kosten [1949], who based their derivation on heuristic differential equations to model the wave propagation in a porous absorber based on phenomenological absorber parameters. Other rigid skeleton approaches based for example on the models by Rayleigh [Strutt, 1945, Vol.II, pp.319-333] or again Zwikker and Kosten [1949] have studied the wave propagation inside a porous absorber on a more microscopic scale, by analyzing the wave motion in a pore of a certain shape considering viscous and thermal effects. Although these approaches present a more rigorous approach to the physical effects inside a porous absorber they inevitably neglect aspects regarding the complex shape and arrangement of pores in typically used fibrous absorbers. More recent semi-phenomenological models have extended these microscopic models to more general micro-geometries. A review of these semi-phenomenological models can for example be found in Allard and Atalla [2009, pp. 73ff]. However, these models necessitate an extensive experimental characterization of the porous absorber material. In order to simplify the experimental characterization of a porous absorber, various empirically based relations for $\underline{\Gamma}_{\mathrm{a}}$ and $\underline{Z}_{\mathrm{a}}$ have emerged, like the Delany and Bazley [1970], Miki [1990] or Komatsu [2008] models. These models use a large set of measured data to establish empirical relations for $\underline{\Gamma}_{\mathrm{a}}$ and $\underline{Z}_{\mathrm{a}}$ as a function of the flow resistivity of an absorber (as well as frequency, density and speed of sound in air).

Although there exist even more complex absorber models like the 'Biot theory' [Biot, 1956], which also accounts for the motion of the absorber skeleton and thus is based on a coupled formulation for the wave propagation of fluid and structural waves in the absorber, the EHF model is often considered adequate to realistically model the energy dissipation in typically used porous absorbers (fibrous or foam absorbers with high porosity). This has different reasons: According to Mechel [1995, p.75ff] (a) the structural waves in typical porous absorbers are generally highly damped due to friction between the fibres and thus only have small amplitudes, (b) the thickness of room acoustic absorbers is often small compared to the structural wave lengths at low frequencies and (c) at high porosities ($> 90\%$) the structural frame has generally a negligible effect on the interaction of a porous absorber with an external fluid sound field, at least if the absorber is not covered with an adhesive airtight foil. Moreover, the 'Biot theory' features a high number of difficult-to-measure model parameters, which necessitates time-consuming and complex measurement setups for the determination of these parameters. These model parameters include the tortuosity of the material, the elastic constants of the absorber skeleton and a factor related to the pore shape. Detailed information on measurement techniques for the determination of these parameters and further references on this topic can be found in Mechel [1995, p.519ff]. Additionally, a complete revision of the different empirical and phenomenological models introduced in this section is given in Mechel [1995] and Allard and Atalla [2009].

In the course of this study the original Zwikker/Kosten model and recent empirical relations by Komatsu were used to calculate the acoustic surface impedances for normal and field incidence on a rigidly backed single layer porous absorber. As a reference acoustic impedances for normal incidence were also measured in the impedance tube at ITA of RWTH Aachen University, which is described in section 5.1. The impedance tube results were averaged over three different samples of the same absorber type and the results for each sample were obtained by averaging four consecutive measurements. In addition to the impedance boundary representation, a 3D absorber model was implemented in the used FE simulation software, which is based on the original model by Zwikker and Kosten. It is thus possible to compare the influence of different models of the same porous absorber on the RTF in a room acoustic FE simulation and also compare these results to measurements conducted in a real room.

7.2.2. Propagation constant and characteristic impedance in the EHF model

- **Zwikker and Kosten model (1949)**

  Since the EHF model neglects the microscopic fine structure of the porous absorber and assumes a fixed structural frame, the only degree of freedom is the mean pressure in the absorber pores. Assuming further that the wave propagation is caused by an interchange between spatially distributed potential and kinetic energy storages in the absorber, Zwikker and Kosten derive the complex propagation constant $\underline{\Gamma}_{\mathrm{a}}$ and the characteristic impedance $\underline{Z}_{\mathrm{a}}$ as follows:

$$\underline{\Gamma}_{\mathrm{a}} = \mathrm{j}k_0\sqrt{\chi\kappa_{\mathrm{eff}}\left(1-\mathrm{j}\frac{\sigma_{\mathrm{v}}\Xi}{\chi\omega\rho_0}\right)} \tag{7.10}$$

$$\underline{Z}_{\mathrm{a}} = \frac{Z_0}{\sigma_{\mathrm{v}}}\sqrt{\frac{\chi}{\kappa_{\mathrm{eff}}}\left(1-\mathrm{j}\frac{\sigma_{\mathrm{v}}\Xi}{\chi\omega\rho_0}\right)}, \tag{7.11}$$

  where the nomenclature of the material parameters is chosen according to that in table 5.2. For a given absorber, the static flow resistivity Ξ and volume porosity σ_{v} are generally determined by straightforward measurement techniques. However, the structure factor χ, which is often considered as a tweaking factor for all the effects that are neglected in the model (typically values range from 1.0 to 1.3), and the corrected adiabatic coefficient κ_{eff} (frequency dependant between 1.41 for low and 1.0 for high frequencies) are usually estimated or simply set equal to 1 (cf. Mechel [1995, p. 97]).

- **Komatsu model (2008)**

  In contrast to the Zwikker/Kosten model, the Komatsu model is an empirically based two parameter model for fibrous sound-absorbing materials, which specifies $\underline{\Gamma}_{\mathrm{a}}$ and $\underline{Z}_{\mathrm{a}}$ as a function of the flow resistivity Ξ and frequency f:

$$\underline{\Gamma}_{\mathrm{a}} = \frac{2\pi f}{c_0}\left[0.0069\left(2-\log_{10}\frac{f}{\Xi}\right)^{4.1} + \mathrm{j}\left(1+0.0004(2-\log_{10}\frac{f}{\Xi})^{6.2}\right)\right] \tag{7.12}$$

$$\underline{Z}_{\mathrm{a}} = \rho_0 c_0\left[1+0.00027\left(2-\log_{10}\frac{f}{\Xi}\right)^{6.2} - \mathrm{j}0.0047\left(2-\log_{10}\frac{f}{\Xi}\right)^{4.1}\right] \tag{7.13}$$

  It has already been mentioned that there exist various more of these empirical relations for porous absorbers, which all yield similar but slightly different absorption characteristics for a given porous absorber (especially in the low frequency range). However, for the sake of brevity we will only include the Komatsu model in this study in order to illustrate the differences between a model based on physically motivated differential equations and an empirical curve fitting model.

7.2.3. Impedances of a porous absorber for normal and field incidence using a 1D transmission line model

Under the assumption that a porous absorber can be described by an EHF model, the acoustic surface impedance $\underline{Z}_{\mathrm{S}}$ of a rigidly backed single layer porous absorber of thickness t is given by the following simple 1D transmission line equation:

$$\underline{Z}_{\mathrm{S}}(\theta_0) = \frac{\underline{Z}_{\mathrm{a}}}{\cos\theta_{\mathrm{a}}}\coth(\underline{\Gamma}_{\mathrm{a}} t\cos\theta_{\mathrm{a}}), \tag{7.14}$$

where θ_0 is the incident angle and θ_{a} is the complex refracted angle inside the porous absorber, which can be calculated from the law of refraction for lossy media

$$\frac{\sin(\theta_{\mathrm{a}})}{\sin(\theta_0)} = \frac{\mathrm{j}k_0}{\underline{\Gamma}_{\mathrm{a}}}. \tag{7.15}$$

As already mentioned, acoustic surface impedances can be assigned as impedance boundary conditions in fluid FEM simulations, but in the case of non-locally reacting room boundary materials the angle of the incident sound wave needs to be a-priori known in order to assign the 'right' impedance to the simulation boundary. Since this is generally not the case, we often confine ourselves to using normal incidence impedances as boundary conditions. However, this simplification is only admissible if the absorber configuration can in good approximation be considered as locally reacting, since in this case $\underline{Z}_S = \underline{Z}_S(0) \neq f(\theta_0)$. For a given absorber the assumption of 'local reaction' can be tested by considering the law of refraction for lossy media (cf. eq. 7.15). The refracted angle is only close to zero for all θ_0 if $|\underline{\Gamma}_{\mathrm{a}}| \gg k_0$, which is generally considered true for porous absorbers with high flow resistivity, due to the high damping term in the propagation constant $\underline{\Gamma}_{\mathrm{a}}$ (cf. eq. 7.10). On the other hand we denote materials with non-negligible wave propagation parallel to the boundary layer as 'laterally reacting'. In the case of porous absorbers such materials generally show increased damping with increasing angle of incidence. This is due to the fact that as the incident angle θ_0 and thus also θ_a increase the actual path length of the sound wave traveling inside the porous absorber is extended and consequently the damping increases.

In order to alleviate the problem of the a-priori unknown angle of incidence, it is possible to adapt an approach which is common practice in GA based simulations. Instead of using normal incidence boundary data in the simulations, boundary data for 'diffuse' or 'field' incidence can be used (cf. fn. 2 in sec. 7.1.4 on the definition of the term field incidence). Similar to the earlier defined field incidence reflection factor, the field incidence surface admittance can be obtained by averaging the surface admittance over all solid angles on a spherical cap with $0° < \theta_0 < 78°$. The corresponding field incidence impedance is then obtained by simple inversion of the admittance term:

$$A_{\text{field}} = \frac{\int_{0°}^{78°} A(\theta)\sin(\theta)\,\mathrm{d}\theta}{\int_{0°}^{78°} \sin(\theta)\,\mathrm{d}\theta} \quad \text{and} \quad \underline{Z}_{\text{field}} = \frac{1}{A_{\text{field}}}. \tag{7.16}$$

Averaging admittances instead of impedances appears admittedly quite arbitrary and it would of course also be possible to obtain a field incidence impedance by directly averaging the angle dependant impedances. Although we could unfortunately not find a good reason to favor the one or the other, it can at least be stated that both methods yield only marginally different results. More importantly, it has to be mentioned that in 2005 Takahashi et al. suggested a different formula for the calculation of the field incidence admittance, which introduces an additional cosine term in the integrals, like it is also found in the Paris formula [Paris, 1927] for the diffuse field absorption coefficient[4]. Again, it was found that the differences between our formula and the formula presented by Takahashi turned out to be rather small. However, from a theoretical point of view we believe that our formula without the cosine term in the integrals is more appropriate for the field incidence admittance. This opinion is based on the fact that the cosine term in the formula for the diffuse field absorption coefficient does not stem from the averaging process of the intensities over the spherical cap but from the fact that the integration is only carried out over the surface normal fractions of sound intensity $I_n = I\cos(\theta_0)$. Using field incidence impedances has the advantage that in the case of laterally reacting materials and field incidence the average damping of the porous absorber is not underestimated as would be the case with normal incidence impedances.

The definitions of $\underline{\Gamma}_a$ and $\underline{Z}_a$ of the Zwikker/Kosten and Komatsu model, which are presented in section 7.2.2, can now be used to calculate the acoustic impedance of a porous absorber for normal and field incidence conditions.

[4]Takahashi simply replaces the angle dependent absorption coefficients in the formula for the diffuse field absorption coefficient (eq. 3.11) by the corresponding admittances:

$$A^*_{\text{field}} = \frac{\int_{0°}^{78°} A(\theta)\sin\theta\cos\theta\,\mathrm{d}\theta}{\int_{0°}^{78°} \sin\theta\cos\theta\,\mathrm{d}\theta}$$

7.2.4. Coupled 3D-FE absorber model according to the Zwikker/Kosten theory

As was already mentioned the 3D-FE porous absorber model used in this study is based on the Zwikker/Kosten EHF model. Since the porous absorber is modeled as a damped fluid with only a pressure degree of freedom the finite element formulation is very similar to that of the fluid domain (see section 4.1.1). A Helmholtz type of equation can thus be derived for the porous absorber as follows:

$$\Delta p - \underline{\Gamma}_{\mathrm{a}}^2 p = 0 \quad \text{and} \quad v = \frac{1}{\underline{Z}_{\mathrm{a}}} \cdot p, \tag{7.17}$$

with the complex propagation constant and characteristic impedance taken from equations (7.10) and (7.11). The corresponding FE formulation for the porous absorber can be derived following the same steps as for the fluid model, where we use the same system matrices $\boldsymbol{K}_F$, $\boldsymbol{M}_F$ and the right-hand side vector $\boldsymbol{f}$:

$$\frac{\mathrm{j}\omega}{\mathrm{j}\omega\frac{\chi}{\sigma_v} + \frac{\Xi}{\rho_0}} \left(\boldsymbol{K}_{\mathrm{F}} + (\mathrm{j}\omega\frac{\sigma_{\mathrm{v}}\Xi}{\rho_0} - \omega^2\chi)\boldsymbol{M}_{\mathrm{F}} \right) \underline{\boldsymbol{p}} = \mathrm{j}\omega\underline{\boldsymbol{f}} \tag{7.18}$$

With regard to the boundary representation in acoustic FE simulations, it is important to point out the differences between an impedance boundary condition, based on the Zwikker/Kosten model for a specific angle of incidence and a coupled 3D-FE absorber model based on the same Zwikker/Kosten theory. It can be said, that the 3D absorber model adjusts the acoustic surface impedance to the right value depending on the actual angle of incidence in the simulation. Thus, in the case where the angle of incidence on a boundary layer is a-priori known and the corresponding impedance is assigned to this boundary, we expect that both models yield the same results for the sound field in the room. In order to show that, preliminary simulations were run in a tube model, which was excited on one side and terminated by a porous absorber on the other side. The simulations were run for frequencies below the cut on frequency of the lowest lateral eigenmode of the tube so that only plane wave propagation was possible in the direction along the tube. The porous absorber was modeled on the one hand with its exact dimensions using the 3D-FE absorber model and on the other hand by an impedance boundary condition which was calculated for normal incidence. As expected the simulation results for the pressure distribution inside the air-filled tube showed a perfect match, which confirms that the underlying theory for the 3D absorber model and the calculation of the impedance boundary condition are the same.

7.2.5. Measurement and simulation setup for the scale model room

The present section summarizes all necessary information on the measurement and simulation setups for the investigated scale model room. This includes a detailed description of the room geometry with its source and receiver positions, the modeling of the boundary materials including the determination of the according input data as well as the characterization of the used sound sources and receivers. Finally, the section discusses the measurements in the real room and the setup for the FE simulations.

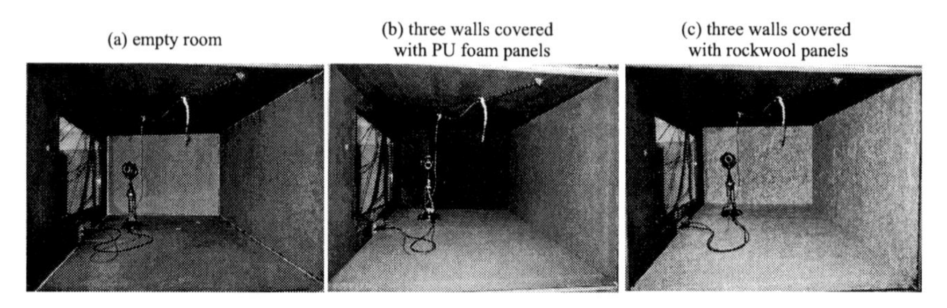

Figure 7.8.: Photographs of the measurement setups in empty condition and with the pu foam and Rockwool panels installed in the scale model room.

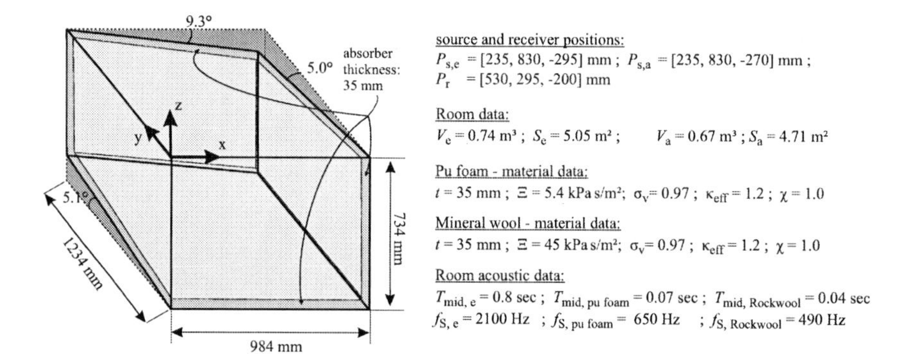

Figure 7.9.: Drawing of the scale model room with chamfered walls and optional absorber mats mounted on three room walls.

P_r = receiver position (for both empty and absorbing condition); $P_{s,e}$ = source position in empty room; $P_{s,a}$ = source position in room with absorber panels (Since the used loudspeaker stand was not adjustable in height the z-coordinate of the source position is slightly higher for the measurements with absorption panel on the floor.). V and S are the room volume and surface with subscripts e and a indicating empty and absorbing condition respectively. The nomenclature of the porous absorber material parameters is chosen according to table 5.2. The mid frequency reverberation time T_{mid} is obtained from the measured third octave band reverberation time T_{20}, which is averaged over the third octave bands from 400 Hz to 1.25 kHz.

Room geometry and source receiver setup

The originally rectangular room is made of 38 mm thick MDF (Medium-Density-Fibreboard) plates and measures $(1.25 \times 1.00 \times 0.75)\,\mathrm{m}^3$. Three walls of the room are planked with 16 mm thick MDF plates that are chamfered at different angles from 5° to 9° in order to avoid flutter echoes in the room. The back sides of the chamfered MDF plates are reinforced with cross-bracings and the cavities between the cross-bracings are filled with mineral wool to avoid strong plate or cavity resonances. Moreover, special cut-to-size porous absorber panels (pu foam and mineral wool) can be mounted optionally on the front sides of the chamfered MDF plates. Figure 7.8 shows photographs of the room in empty condition and with the pu foam and Rockwool panels installed in the room. The photographs also show the source and receiver configuration that was used throughout this study. An additional drawing of the CAD model of the room with all relevant dimensions, material properties and the exact source and receiver positions is given in Figure 7.9. The dimensions, positions and angles given in the plot were carefully measured in the real room and then used to generate the CAD model and FE mesh of the room.

Boundary materials in simulation and measurement

- **MDF Walls**

  It is clear that the thick MDF walls have an almost rigid reflection characteristic. Thus in a room with other dominant absorbers it might be sufficient to consider these walls as perfectly rigid in an FE simulation. However, in the case of the empty room or generally in frequency ranges with low average absorption an estimation of the impedance characteristics of the MDF walls is inevitable to realistically capture the small energy dissipation on the walls. Moreover, this determination needs to be very precise since in an almost undamped room the reverberation time is very sensitive to slight changes in the wall absorption; e.g. a small change in the absolute average wall absorption from 0.04 to 0.06 leads to an approximate change in the reverberation time of 50% (according to the Sabine reverberation law).

  Since an exact determination of the impedance characteristic of an almost rigid material appears very difficult with the impedance measurement techniques introduced in section 5.1-5.3, we have calculated a homogeneous, real-valued, impedance (in third octave bands) on the basis of the reverberation times obtained from the empty room measurements. Despite the fact that one of the MDF walls has a glass window in it, we have, for the sake of simplicity, assumed that all walls in the empty room have the same reflection characteristics. In particular, we used the Eyring reverberation law (eq. 3.24) to calculate the average diffuse field absorption coefficient of the room walls and then determined the real valued surface impedance using equation 5.2. This impedance was then homogeneously assigned to all MDF walls (including the window) in the simulations of the scale model room. Unfortunately, as will be discussed in detail in section 7.2.6, it was found, that the empty room simulation using the so determined impedance values led to considerably higher reverberation times than were obtained from the corresponding measured room transfer function. This discrepancy is most likely due to insufficient diffusion in the scale model room, which makes the application of the Eyring reverberation law and thus equation 5.2 surely questionable (for more details cf. section 5.4). In order to get a better match between simulated and measured reverberation times in the empty rooms, the real valued impedances have therefore been adjusted by using the following simple iterative algorithm:

$$\alpha_{\mathrm{sim,k+1}} = \alpha_{\mathrm{sim,k}} \cdot \frac{T_{\mathrm{sim,k}}}{T_{\mathrm{meas}}}, \tag{7.19}$$

$$\text{with} \quad \alpha_{\mathrm{sim,k}} = 1 - \left|\frac{Z_{\mathrm{sim,k}} - Z_0}{Z_{\mathrm{sim,k}} + Z_0}\right|^2 \tag{7.20}$$

$$\text{and} \quad Z_{\mathrm{sim,k+1}} = Z_0 \cdot \frac{1 + \sqrt{1 - \alpha_{\mathrm{sim,k+1}}}}{1 - \sqrt{1 - \alpha_{\mathrm{sim,k+1}}}}, \tag{7.21}$$

where k is the iteration step, and T is the reverberation time. A comparison of the results obtained with the adjusted and non-adjusted impedances are reported in section 7.2.6. The adjusted real-valued impedance values were then also assigned to the non-covered room walls in the simulations of the damped room.

- **Rigidly Backed Porous Absorber Panels**

  For the investigations in the damped scale model room two different porous materials with largely different flow resistivities were used in order to emphasize the differences between locally and laterally reflecting materials. The first material is a typical porous pu foam, with thickness $d = 35$ mm, and measured parameters $\Xi = 5.4\,\mathrm{kPa\,s/m^2}$ and $\sigma_{\mathrm{v}} \approx 0.97$. The second material is a Rockwool mineral wool of the same thickness with measured parameters $\Xi = 45\,\mathrm{kPa\,s/m^2}$ and $\sigma_{\mathrm{v}} \approx 0.97$. In addition to the measured absorber parameters reasonable estimated values of $\chi = 1$ and $\kappa_{\mathrm{eff}} = 1.2$ were used for both materials[5]. In the simulations these absorbers were modeled as rigidly backed single layer absorber panels using the models described in section 7.2.2.

  Figure 7.10 shows a comparison of the measured (impedance tube) and calculated (Zwikker/Kosten and Komatsu) absorption coefficients for normal incidence for both materials. Despite the use of complex impedances as boundary conditions in the FE calculations, we plot absorption coefficients to facilitate the interpretation of the results. For the pu foam the Komatsu model gives a good match with the measured absorption coefficients, whereas the Zwikker/Kosten model yields on average about 0.15 less absorption. For the Rockwool mineral wool both models show a reasonable match with the measured data.

[5] Although from a theoretical point of view a value of $\kappa_{\mathrm{eff}} = 1.41$ would have been be more suitable, due to the rather low frequencies in the FE simulations, the value was adjusted to 1.2 for better agreement with the measured absorption data. However, it has to be mentioned that the influence of different values of κ_{eff} on the absorption data is rather small.

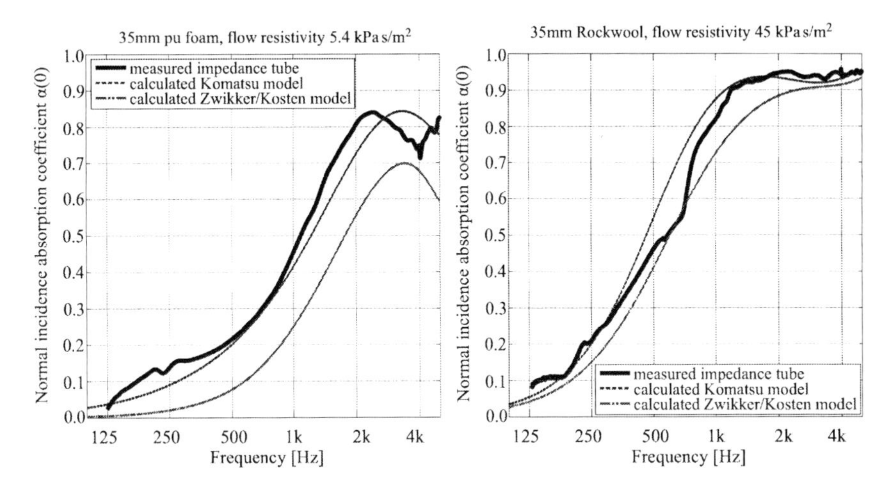

Figure 7.10.: Absorption of pu foam and Rockwool mineral wool for normal incidence calculated from measured and modeled impedances

Figures 7.11 and 7.12 show a contour plot of the absorption coefficient and a comparison of the normal incidence and field incidence impedance curves for both materials. These curves were calculated using the Komatsu model which showed on average the best match with the measured data from the impedance tube. The respective curves calculated with the Zwikker/Kosten model yield qualitatively the same results and are therefore not reported here. As expected, it can be seen that the pu foam with low flow resistivity shows a much stronger angular dependency of the absorption coefficient than the mineral wool with high flow resistivity. However, it is important to mention that only looking at absorption curves can be somehow misleading in this context, since the absorption coefficient always shows an angular dependency due to the change in the characteristic impedance of air, $Z_0' = \frac{Z_0}{cos(\theta_0)}$, even for perfectly locally reacting materials. Consequently, an angular dependence of the absorption coefficient does by itself not indicate an angular dependence of $\underline{Z}_S$ and thus a 'non-locally reacting' material. Looking at the results in figure 7.12 indeed indicates that only in the case of the pu foam the angular dependence of α is strongly driven by an angular dependence of $\underline{Z}_S$ whereas in the case of the Rockwool absorber the slight changes in α are mostly caused by the change in Z_0'. In section 7.2.6 it will be discussed how the laterally reacting characteristics of the pu foam affect the simulation results when an impedance boundary condition is used and it will be investigated if improved results can be obtained with the 3D absorber model, which fully captures the angular dependency of the acoustic impedance of the absorber.

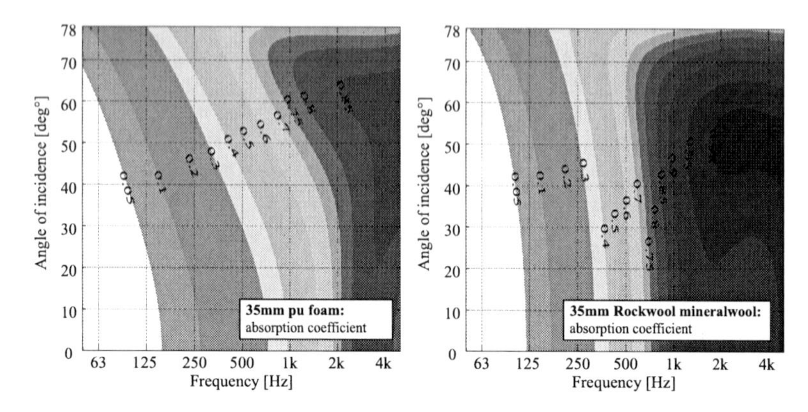

Figure 7.11.: Contour Plot of the absorption coefficient of pu foam and Rockwool mineral wool for different angles of incidence calculated from the Komatsu model.

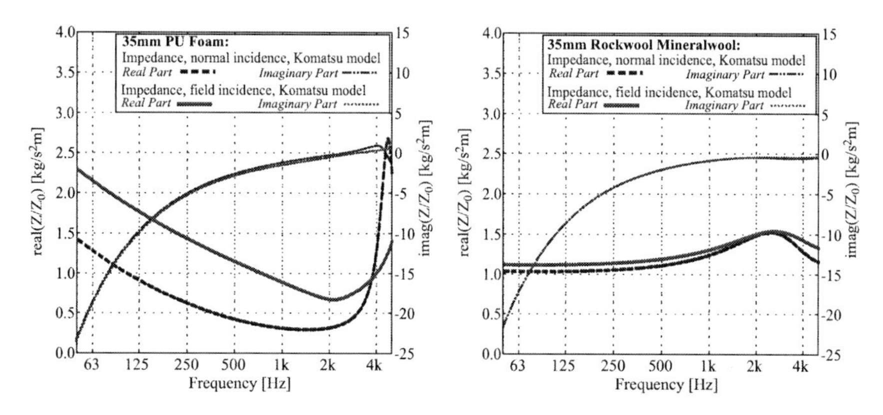

Figure 7.12.: Comparison of normal incidence and field incidence acoustic surface impedance for pu foam and Rockwool mineral wool calculated from the Komatsu model.

Sound source and receiver in simulation and measurement

For the room acoustic measurements in the scale model room two different loudspeakers and microphones were used to assess the uncertainty in the measured low frequency room transfer functions caused by different measurement equipment.

As sound sources a small dodecahedron and a small hexahedron loudspeaker were used, which were both developed at ITA of RWTH Aachen University and which are shown in figure 7.13. In order to account for the different frequency characteristics of the used loudspeakers, anechoic measurements of the on-axis free field pressure response and directivity

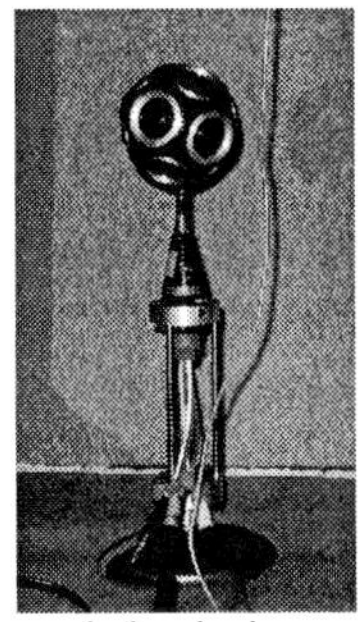

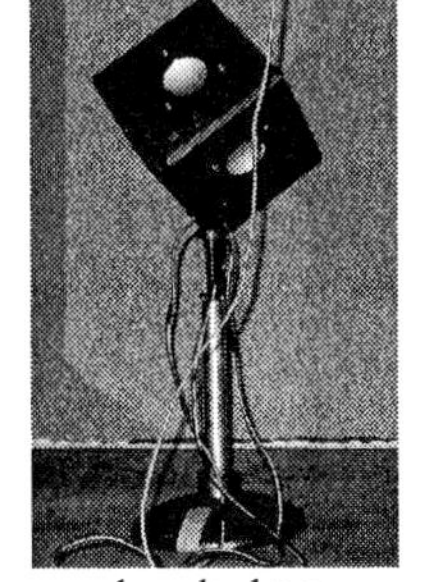

Figure 7.13.: The small dodecahedron and hexahedron loudspeaker, which were used for the room acoustic measurements in the scale model room.

(cf. section 6.1.1) were conducted for both loudspeakers, where the free field pressure responses were normalized to 1 m distance and 1 V input voltage at the loudspeaker input. Taking into account that both loudspeakers showed an almost omnidirectional characteristic up to at least 1.5 kHz, the room acoustic measurements in the scale mode room, which were also conducted re 1 V at the loudspeaker input, were normalized by a subdivision with the magnitude of the averaged free field pressure response of the respective loudspeaker. Consequently, the applied normalization leads to loudspeaker independent low frequency RTFs that are obtained for an omnidirectional sound source with a frequency constant sound pressure of 1 Pa at 1 m distance in the free field. In the FE simulation such a sound source is adequately modeled by a simple monopole point source with a normalized source strength of $A = 1\,\mathrm{kg}/\mathrm{s}^2$, where the relation between the volume velocity Q as introduced in equation 3.7 and the source descriptor A, which is for example used in *Virtual Lab*, is given by $Q = \frac{4\pi}{\mathrm{j}\omega\rho_0} \cdot A$.

As measurement microphones a 1/2" *Brüel & Kjær* microphone of type 4190 and a Sennheiser KE-4 capsule were used. In order to obtain absolute sound pressure levels from our measurements the input measurement chain was fully calibrated for each microphone using a pistonphone with an excitation frequency of 1 kHz. Since both microphones were expected to have an almost flat frequency response in the considered frequency range from 80 Hz to 1.5 kHz they were considered as ideal point receivers in our simulations.

It should be emphasized again, that the described choice of sound sources and microphones together with a full calibration of the input and output measurement chain allows a direct comparison of measured and simulated room transfer functions without any further adjustment. Moreover, the used sound sources and receivers in the measurements were deliberately chosen such that they correspond as good as possible to their idealized counterparts in the simulation. This is obviously crucial when discussing the accuracy of different boundary models on the basis of a comparison of measured and simulated room transfer functions.

Measurements in the real room

As was already mentioned in the previous section the measurements in the scale model room were carried out with different loudspeakers and microphones to assess their influence on the uncertainty of the measured low frequency responses of the room. In addition to the use of different measurement equipment, the measurement results were also obtained in two completely independent measurement sessions in 2009 and 2011, in which the room setup as well as the source and receiver positions used in the FE model were replicated to the best possible extend. In both measurement sessions the exact same scale model room and absorber panels were used. However, the installation of the chamfered MDF wedges, the mounting of the absorber panels and the positioning of the source loudspeaker and the measurement microphone in the room were redone completely from scratch in both sessions. While in the session in 2011 both the dodecahedron and the hexahedron loudspeaker were used in combination with the KE-4 capsule, the session in 2009 was conducted using only the small dodecahedron loudspeaker and the *Brüel & Kjær* microphone. Thus, three different measurement runs can be compared, which are denoted "2009 - Dode to B&K", "2011 - Dode to KE-4" and "2011 - Hexa to KE-4". All three measurement runs were carried out for the empty room condition, the room with the pu foam panels and the room with the Rockwool panels mounted to the walls.

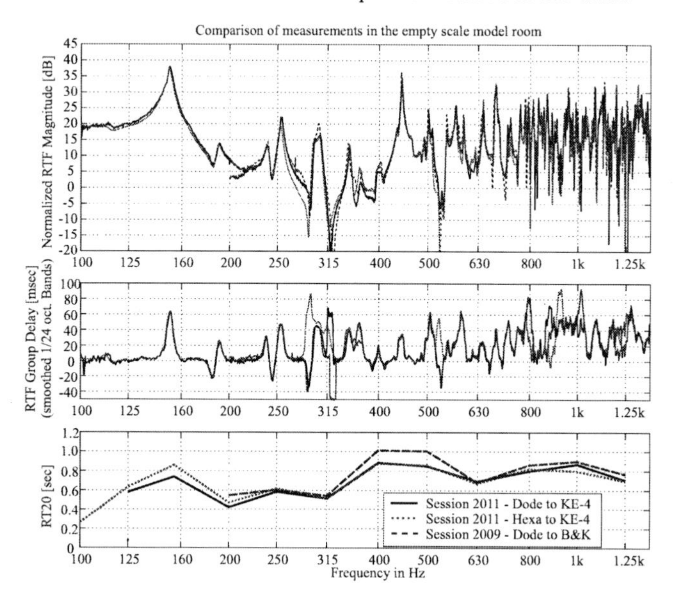

Figure 7.14.: Comparison of measured room transfer functions and reverberation times in empty room condition.
The measurement results were obtained in two independent measurement sessions in 2009 and 2011. The results for '*Session 2009 - Dode to B&K*' are only plotted for frequencies above 200 Hz since the used measurement sweep only covered the frequency range above that frequency. The measured room transfer functions are normalized to the sound pressure in 1 m distance in the free field generated by the respective source.

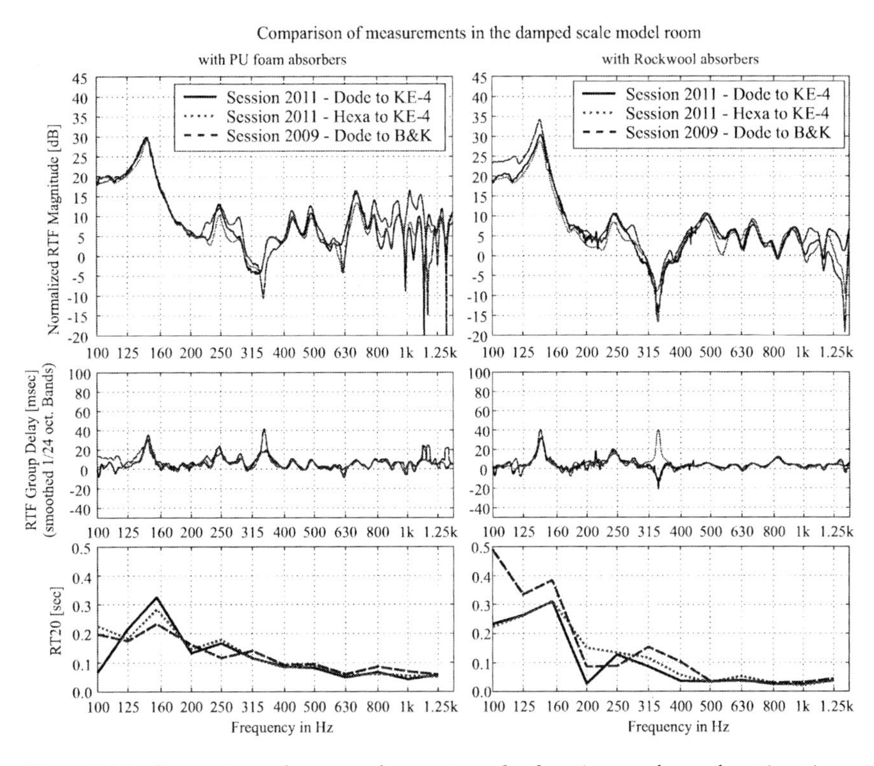

Figure 7.15.: Comparison of measured room transfer functions and reverberation times in the damped room with PU foam panels (left plot) and Rockwool panels (right plot) mounted on three of the room walls.
The measurement results were obtained in two independent measurement sessions in 2009 and 2011. The measured room transfer functions are normalized to the sound pressure in 1 m distance in the free field generated by the respective source.

Figures 7.14 and 7.15 show a comparison of the RTF magnitudes, the RTF group delays[6] (smoothed in $\frac{1}{24}$ octave bands for better readability) and the reverberation times obtained in the three measurement runs. The measurement results show an overall good agreement with regard to the magnitude and group delay of the low frequency transfer functions as well as the reverberation times. However, considerable differences in the fine structure of the room transfer function can be observed for the room with the Rockwool absorber panels (above 1 kHz) and for the room with the pu foam panels (between 200 and 400 kHz as well as above 800 Hz). Except for the low frequency deviations in the room with pu foam panels these differences occur predominantly in frequency ranges above the Schroeder

[6]The group delay τ_g is given in seconds and is defined as the rate of change of the total phase shift ϕ with respect to the angular frequency ω: $\tau_g = -\frac{d\phi}{d\omega}$. We prefer plotting group delay over plotting the unwrapped phase because of the smaller range of values in the group delay plot.

frequency of the considered room. It is thus most likely that these discrepancies are caused by the sum of many small inevitable differences between the measurement setups. These include positioning inaccuracies of the source and receiver, the influence of the different measurement equipment and differences in the mounting conditions of the absorber panels (caused by thin air-gaps at the edges or behind the absorber panels). While it is surely not the aim of the study to immerse into the wide field of uncertainty in room acoustic measurements, it is important to bear in mind that the agreement of measured and simulated room transfer functions should always be judged on the basis of the uncertainty of the measured results. A simulation result is thus hardly improvable if the deviations between the simulated and a measured result lie in the same order of magnitude as the deviations between repeated measurement results of the same scenario.

When comparing measured and simulated room transfer functions in section 7.2.6, we only plot the measurement result that yields the best match with the respective FE simulation for a better readability of the plots.

FE simulation setup and room model

With regard to the FE simulation setup the fundamental requirements specified in section 4.1.3 were satisfied to the best possible extend. In order to achieve this, the CAD model of the room was meshed in compliance with equation 4.5 for an upper frequency limit of $f_{\mathrm{max}} \approx 2\,\mathrm{kHz}$ (average edge length of $\approx 50\,\mathrm{mm}$). All FE simulation were run in the frequency range from 80 Hz to 1.6 kHz. Additionally the frequency step width has been chosen in consideration of equation 4.9 with $\Delta f_{\mathrm{e}} = 0.5\,\mathrm{Hz}$ for the empty room and $\Delta f_{\mathrm{a}} = 0.5\,\mathrm{Hz}$ between 80-200 Hz, $\Delta f_{\mathrm{a}} = 1\,\mathrm{Hz}$ between 201-400 Hz and $\Delta f_{\mathrm{a}} = 2\,\mathrm{Hz}$ between 402-1600 Hz for the room with porous absorption. The coarsening of the frequency step width at higher frequencies in the simulations of the damped room is justified on the basis of the steadily decreasing reverberation times with increasing frequency. The source and receiver positions were chosen according to figure 7.9. In the absorbing case where the porous panels were modeled by an acoustic surface impedance, the reduction in the fluid volume caused by the thickness of the absorber panels was accounted for by cutting out the absorber volumes from the fluid volume. Likewise, in case of the 3D porous absorber model, the volume of the absorber panels was defined as a porous absorber domain according to section 7.2.4 and coupled to the fluid domain inside the room.

7.2.6. Objective comparison of measurement and simulation results

The present section compares the measurement and simulation results obtained for the empty room and the room with porous absorber panels applied to three room walls and discusses the reasons for the observed differences between the different boundary models.

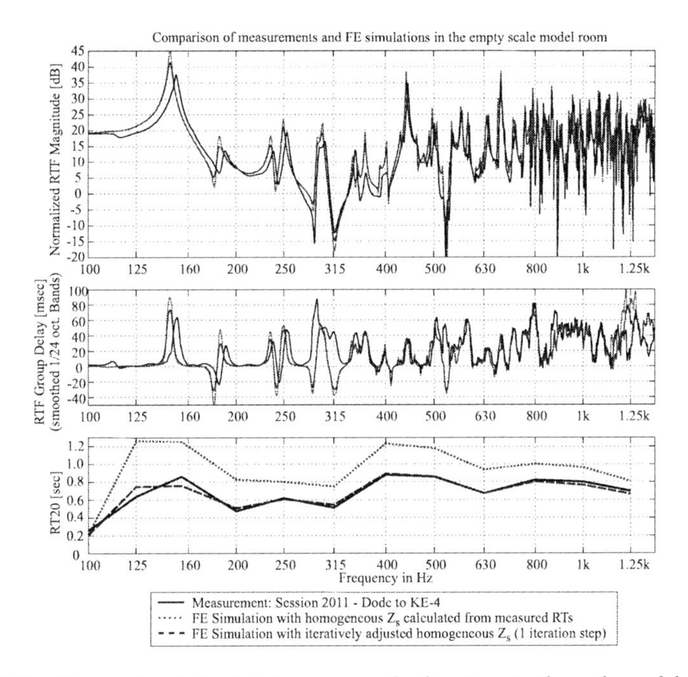

Figure 7.16.: Measured and simulated room transfer functions in the scale model room in empty condition
Measured and simulated room transfer functions are normalized to the sound pressure in 1 m distance in the free field generated by the respective source.

Empty room

A comparison of the normalized measurement and simulation results for the source receiver pair $(P_{s,e}, P_a)$ in the empty room is given in figure 7.16. As was already mentioned above the FE simulations were run with different wall impedances assigned to the hard and reflective MDF walls in the room. In the first simulation a real-valued third octave band impedance, which was calculated from the measured reverberation times by using equations 3.24 and 5.2, was homogeneously assigned to all room walls. As can be seen from the reverberation time plot in figure 7.16 the so determined impedance values consistently underestimate the damping in the empty room throughout the considered frequency range. This result clearly shows that the sound field assumptions underlying the Eyring reverberation law do not hold in the considered room and frequency range. This is surely not surprising, considering the lack of diffusing elements and the fact that the simulated frequency range lies well below the Schroeder frequency. However, since we could not find a better direct way to deduce more accurate impedance values of the hard reflective MDF walls, we have resorted to the iterative algorithm described in equation 7.19 to determine more realistic average surface impedances. As can be seen from figure 7.16 only one iteration sufficed to obtain a very good match of the measured and simulated reverberation times.

Irrespective of the reverberation times that were obtained from the FE simulations, the FE results show an overall very good agreement of the measured and simulated RTF with regard to the logarithmic amplitude and the smoothed group delay up to frequencies higher than 1 kHz. The difference in the overall damping of the room between the two FE simulations is only indicated by the higher modal quality factors in the simulation with the impedances deduced from the measured RTs. On the other hand, the observed slight mismatch between measured and simulated eigenmode frequencies in the lowest frequency range, might be due to small inaccuracies in the geometrical model or in the assumed values for the sound velocity or air density ($c = 344\,\mathrm{m/s}, \rho_0 = 1.205\,\mathrm{kg/m^3}$).

The overall very good match between measured and simulated data indicates that the used room CAD model, the normalized point source model and the iteratively determined real-valued surface impedances for the MDF walls constitute an adequate model for the simulation of the empty room measurement setup. We are thus confident that the observed differences between measurement and simulation in the damped scale model room, which will be discussed in the following section, are predominantly caused by the representation of the porous absorber materials at the room boundaries.

Damped room

We first want to discuss the effect of angular dependence of the acoustic impedance on the simulated sound field in the considered room. Different FE simulations have therefore been conducted, where the acoustic characteristics of the absorbing room boundaries are either considered by an acoustic impedance calculated for normal incidence or a 3D-FE absorber model, both of which are derived on the basis of the Zwikker/Kosten EHF model. The simulations were run for the pu foam and the Rockwool mineral wool. Additionally, as a reference, measurements in the scale room have been conducted where three walls were covered with the pu foam and the Rockwool mineral wool respectively. Since the two different boundary representations are all based on exactly the same absorber model, it is interesting to ask the question in which cases different results are expected. Figure 7.17 shows the logarithmic amplitude, the group delay and the deduced reverberation time of the resulting normalized RTFs. It can be seen that for the mineral wool the two boundary representations yield almost identical results which show a reasonably good match with the measured curve. This is due to the fact that the mineral wool shows almost no angular dependence of the acoustic impedance in the considered frequency range (see figure 7.12). In other words this means, that due to its high flow resistivity the mineral wool absorber can be classified as almost perfectly 'locally reacting'. Since all incident waves are refracted to the normal direction inside the porous absorber there is no benefit from the 3D-FE absorber model, since the 1D transmission line model (used to calculate the surface impedance for normal incidence) fully captures the acoustic behaviour of this absorber. The calculated reverberation times also show an overall good match with the measured RTs, although a slight underestimation of the damping in the room can be observed for the simulation results.

In case of the pu foam things look different. Due to the non-negligible angular dependence of the acoustic characteristics of the pu foam, the results obtained with the impedance for normal incidence show considerably higher quality factors and thus lower damping of

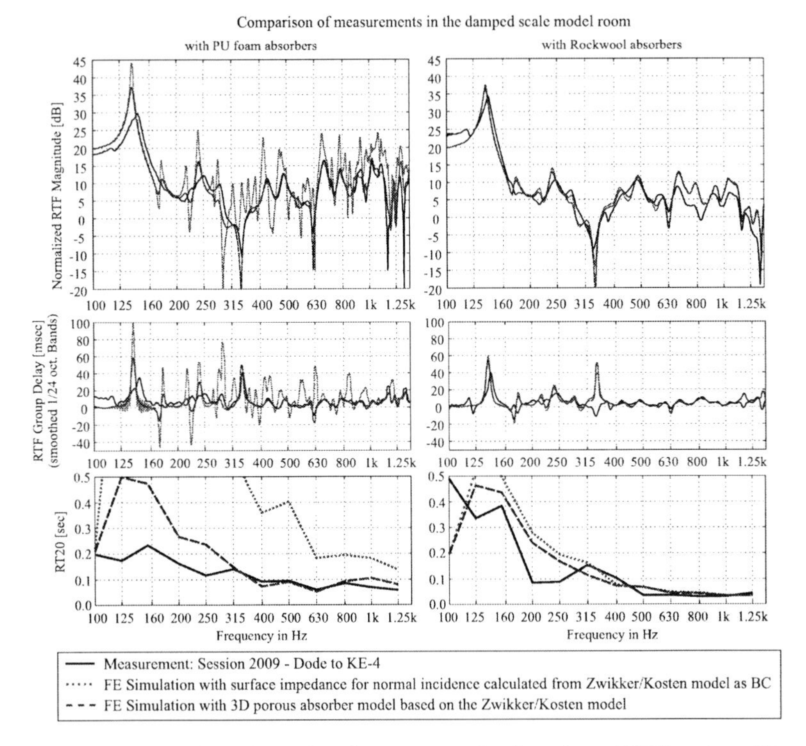

Figure 7.17.: Measured and simulated room transfer functions (RTF) using different boundary models for porous absorber (1).
Measured and simulated room transfer functions are normalized to the sound pressure in 1 m distance in the free field generated by the respective source.

the room eigenmodes than is obtained with the 3D-FE absorber model. When compared to the measurements in the model room, it is found that the 3D absorber model yields very good results, whereas the normal incidence impedance strongly underestimates the damping in the room. It can thus be concluded that in the case of 'non-locally reacting' absorbers, a normal incidence acoustic impedance does generally underestimate the absorption characteristics of a porous absorbing room boundary, as was expected from our explanations in section 7.2.3.

The reverberation times obtained from the simulated and measured room transfer functions with the pu foam panels corroborate the discussed results. In particular, the simulation with the normal incidence impedance shows a strong underestimation of the damping in the room. On the other hand, the observed too high reverberation times for the simulations with the 3D absorber model for frequencies below 315 Hz might be explained by the too low absorption values obtained from the Zwikker/Kosten model when compared to measured absorption values in the impedance tube. Moreover, at very low frequencies

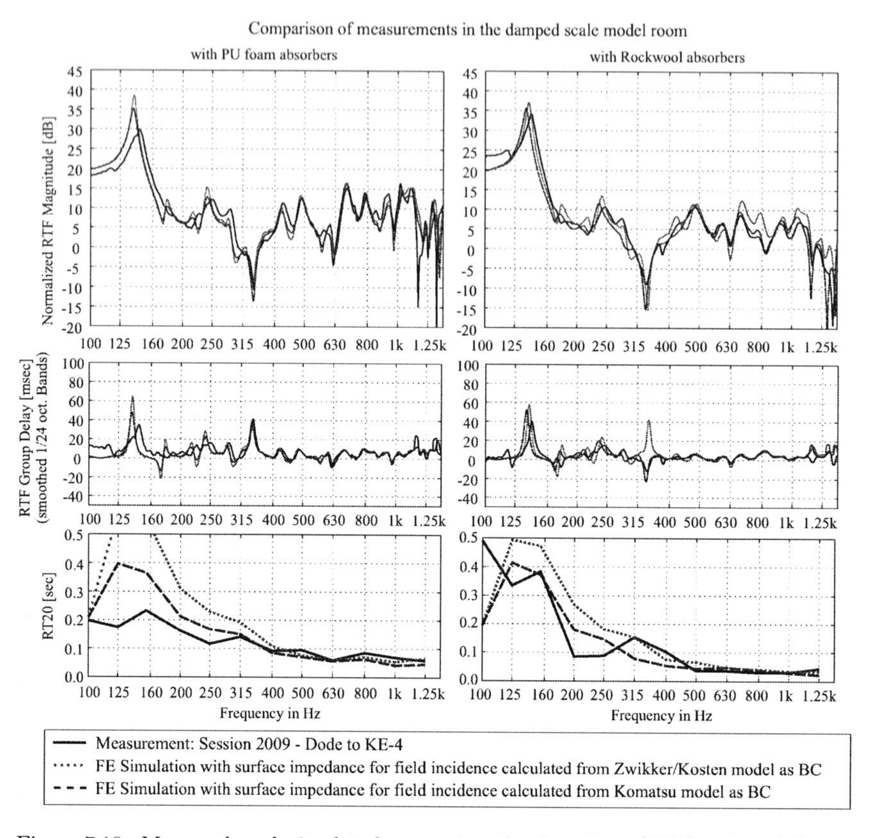

Figure 7.18.: Measured and simulated room transfer functions (RTF) using different boundary models for the porous absorber (2).
Measured and simulated room transfer functions are normalized to the sound pressure in 1 m distance in the free field generated by the respective source.

other damping mechanisms and energy losses which are not considered in the simulations might also play a role; e.g. leakage at the edges of the chamfered MDF plates (which are sealed with modeling clay) or leakage at the rubber seal at the edges of the front wall of the room (the front wall plate can be detached to access the interior of the scale model room).

In the next step the performance of the field incidence impedance model shall be investigated. Figure 7.18 shows the results of these simulations, where the Zwikker/Kosten model and the Komatsu model were used to calculate the field impedances of the pu foam and mineral wool respectively. For the pu foam, it can be seen that the RTFs calculated with the field incidence impedances as boundary conditions yield a very good match with the measured RTF. However, although the absorption curves of the pu foam in figure 7.10 show that the Komatsu model yields a better match with the measured data than is obtained from the Zwikker/Kosten model, a considerable improvement in the match of measured and simulated data can only be found in the reverberation time plots. It is also noteworthy that for the pu foam the field incidence impedance model using the Zwikker/Kosten model yields comparably good results for the RTF as those obtained with the full 3D absorber model.

In the case of the mineral wool it is found that using the field incidence impedance calculated from the Komatsu model yields a slightly better match with the measured RTF (especially in the range from 600 Hz - 1 kHz) and RT characteristics (at frequencies below 200 Hz) than the respective impedance calculated from the Zwikker/Kosten model, which gives almost identical results to those obtained from the normal incidence impedance or the 3D-FE absorber model based on this model. Only at around 345 Hz a significant difference between both simulations is observed in the group delay. Taking into account that two very close eigenfrequencies of the rigid room are found in this frequency range and further considering the sharp notch in the magnitude of the RTFs at about 345 Hz, it is believed that this difference is due to a strong resonant effect caused by a slight discrepancy in the room eigenmodes simulated with the different absorber models on the boundaries. Thus, even in the case of perfectly 'locally reacting' materials, where a simple normal incidence impedance suffices, the simulation results obviously depend on the applied absorber model and the corresponding impedance. The uncertainty in the RTF which is obtained by using different porous absorber models appears to be in a range of on average 1 to 2 dB.

7.2.7. Summary

Summing up, the results of the study confirm the expectations that normal incidence acoustic impedances are only valid as boundary conditions in room acoustic FE simulations, if the considered porous absorber is in good approximation 'locally' reacting. Moreover the results show that in the case of 'laterally' reacting porous absorbers a normal incidence impedance generally leads to a considerable underestimation of the damping in a room, while the 3D-FE absorber model and the field incidence impedance show considerably improved simulation results, since these models capture the increased damping that is obtained for angular incidence. Interestingly, although the field incidence impedance does only account for the angle of incidence on an average basis, the simulations with field incidence impedances show comparably good results to those obtained with the 3D-FE absorber model. Taking into account the higher computational cost and model complexity associated with a 3D-FE absorber model, it is concluded that a field incidence impedance adequately captures the characteristics of both 'locally' and 'laterally' reacting porous absorbers in room acoustic FE simulations and that it is possible to obtain an excellent agreement between measured and simulated results with this approach. However, it has to be mentioned that the simulation results obviously depend on the underlying porous absorber model and the quality of its input data.

7.3. Coupling of low frequency room sound field and loudspeaker diaphragm velocity

The following section discusses the question raised in section 6.1 regarding the possible coupling of the low frequency sound field in small rooms and the loudspeaker diaphragm velocity. The presented investigations are based on an FE simulation study, in which three different simplified models of acoustically relevant small spaces (office room, car passenger compartment, loudspeaker enclosure) are used with a typical low- to midrange loudspeaker source, which is characterized by its Thiele-Small parameterization. Although the presented results are of course dependent on the chosen loudspeaker and room types, we believe that the conclusions from this simulation study can be generalized and also apply to other loudspeakers and rooms.

7.3.1. Loudspeaker and room data

The present section gives some details on the used loudspeaker and room models.

1. The used low- to midrange loudspeaker has the following Thiele-Small parameterization:

   $B \cdot l = 3.90\,\mathrm{N/A}$; $S_\mathrm{d} = 143.14\,\mathrm{cm}^2$; $R_\mathrm{e} = 1.84\,\Omega$; $L_\mathrm{e} = 0.102\,\mathrm{mH}$; $R_2 = 4.94\,\Omega$; $L_2 = 0.216\,\mathrm{mH}$; $R_\mathrm{s} = 1.284\,\mathrm{kg/s}$; $M_\mathrm{d} = 15.287\,\mathrm{g}$; $C_\mathrm{s} = 0.283\,\mathrm{mm/N}$.

   Since the loudspeaker box is assumed to be closed in all simulations the elements representing the vent are omitted. For the office room and car compartment simulations the closed loudspeaker enclosure is approximated by a simple compliance: $C'_{ab} = \frac{V_0}{cZ_0}$, with $V_0 = 0.0108\,\mathrm{m}^3$; $M'_\mathrm{ab} = 0\,\mathrm{kg/m^4}$ and $R'_\mathrm{ab} = 0\,\mathrm{kg/sm^4}$ and the radiation impedance $\underline{Z}_\mathrm{rad}$ is obtained from the FE simulations in the considered room. On the other hand, in the simulations for the loudspeaker enclosure, the impedance terms related to the loudspeaker box (C'_ab, M'_ab and R'_ab) are replaced by the corresponding impedance obtained from the FE simulations and the front side radiation impedance $\underline{Z}_\mathrm{rad}$ is set to the free field radiation impedance of a piston in an infinite baffle (cf. eq.: 7.22).

2. In order to investigate the influence of the low frequency sound field in an enclosed cavity on the diaphragm velocity, three different exemplary small spaces were used, where each space was simulated for a low and moderate damping setting. The corresponding room data is summarized in figure 7.19.

7.3.2. Theoretical considerations

To approach the above raised question we can state ex ante that (a) an influence of the room sound field on the loudspeaker behaviour (i.e. on the membrane velocity) is only expected if the radiation impedance on the loudspeaker diaphragm is not small compared to the other impedances in the acoustical loudspeaker circuit. If this is the case it has to

	Office room	Car passenger compartment	Loudspeaker box
Volume [m^3]	5x4x3 = 60	2.55	0.2x0.18x0.3 = 0.0108
Surface [m^2]	94	17.2	0.3
Z_S/Z_0 [-]	1.055 + 8j ; 4.5 - 6j	1.055 + 8j ; 4.5 - 6j	∞ ; 71 + 0j ; por.Abs.*
α_{diff} [-]	0.1 ; 0.4	0.1 ; 0.4	0.0 ; 0.1 ; por.Abs.*
T_{Sabine} [s]	1.04 ; 0.26	0.24 ; 0.06	-
$f_{\text{Schroeder}}$ [Hz]	263 ; 132	614 ; 307	-

Figure 7.19.: Summary of geometrical and acoustical room/cavity data.
The loudspeaker membrane is modeled as a piston which is highlighted in dark brown in the CAD models of the rooms/cavities. A homogeneous surface impedance Z_S is assigned to all boundary surfaces except for the piston surface.
* For this simulation the inner walls of the loudspeaker box were considered rigid, but the interior was filled with a porous absorber according to the model by Zwikker and Kosten [1949] with $\Xi = 5\,\text{kPa s/m}^2$, $\sigma_v = 0.98$, $\chi = 1.3$, $\kappa_{\text{eff}} = 1.4$ (cf. section 5.3 and 7.2.2 for more information on this 3D absorber model.)

be further investigated if (b) the radiation impedance considerably varies depending on the source position, room size or absorption characteristics on the room boundaries. To start with and as a reference we want to consider the radiation impedance of a loudspeaker in the free field which in the case of a conical loudspeaker membrane can be approximated by considering a piston of surface S_d with constant velocity v_d in an infinite baffle. In this idealized case the radiation impedance is given by [Mechel, 1989, p.324]:

$$\underline{Z}_{\text{rad,ff}}(f) = \rho_0 c \left(1 - \frac{J_1(2ka)}{ka} + j\frac{S_1(2ka)}{ka}\right), \quad \text{with} \quad a = \sqrt{\frac{S_d}{\pi}} \tag{7.22}$$

where $J_1(z)$ and $S_1(z)$ are the Bessel and Struve functions of first order and a is the radius of the piston area. With regard to question (a) figure 7.20 shows this free field radiation impedance in comparison to the sum of the other components in the acoustical network for the considered loudspeaker. It can be seen that throughout the whole audible frequency range the impedance contribution from the radiation impedance is at least 20 dB below that of the summed mechanical and electrical part of the closed box loudspeaker circuit. Additionally, figure 7.21 compares the membrane velocities of the loudspeaker for $\underline{Z}_{\text{ar}} = \underline{Z}_{\text{rad,ff}}/S_d$ and $\underline{Z}_{\text{ar}} = 0$ (which corresponds to a vacuum on the front side of the loudspeaker diaphragm). This comparison also corroborates that for a loudspeaker operating in the free field the contribution of the radiation impedance to the diaphragm velocity can generally be neglected. This result is well accepted in the literature (e.g. Thiele [1971a, p.384]).

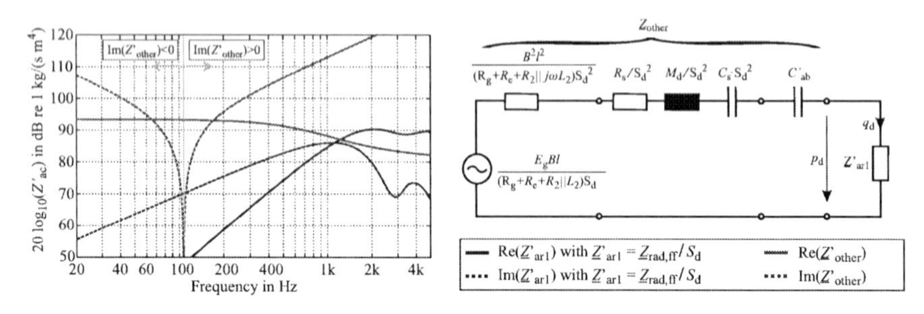

Figure 7.20.: Left: Comparison between acoustic free field radiation impedance vs. summed impedance contributions from electrical part, mechanical part and loudspeaker box compliance for typical low- to midrange loudspeaker. Right: Loudspeaker circuit diagram according to figure 6.2 with omitted vent part and all components transformed to the acoustical side.

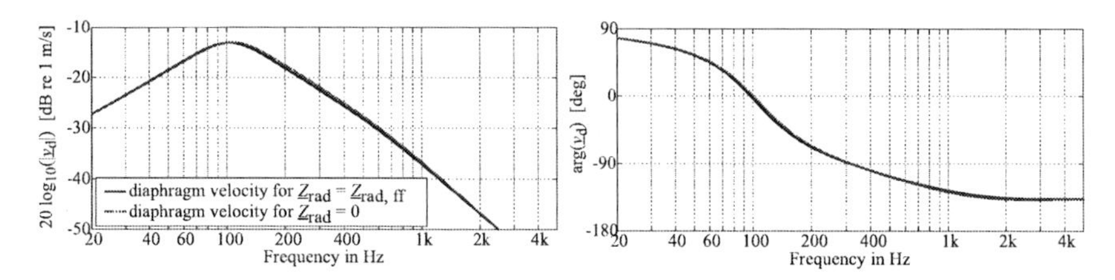

Figure 7.21.: Diaphragm velocity of closed boxed loudspeaker with $\underline{Z}_{\text{ar}} = \underline{Z}_{\text{rad,piston}}/S_{\text{d}}$ and $\underline{Z}_{\text{ar}} = 0$ (corresponds to vacuum on the front side of the diaphragm).

7.3.3. Simulation results

In a next step, we now want to investigate if the radiation impedance considerably changes if the loudspeaker operates into a closed air-filled cavity instead of the free field. Figures 7.22 to 7.25 show the radiation impedance and the diaphragm velocity (for $E_{\text{g}} = 1\,\text{V}$) obtained from FEM simulations in the office room and in the car compartment where the piston surface is excited with $v_{\text{n}} = 1\,\text{m/s}$ and the sound pressure is averaged over the piston area. It can be seen, that in the case of the medium sized office room the radiation impedance is very similar to that obtained in the free field case and consequently the membrane velocity obtained with the radiation impedance from the FEM simulation in the room hardly differs from that obtained with the free field radiation impedance $\underline{Z}_{\text{ar,piston}}$. In the case of the simplified car model things look different. Especially in the low absorbing case (with $\alpha_{\text{diff}} = 0.1$) the strong modal structure of the low frequency sound field can also be found in the radiation impedance. However, even for this strong deviation from the free field radiation impedance only a very weak influence is exerted on the diaphragm velocity (cf. figure 7.25). For a more realistic average damping in the car passenger compartment (α_{diff}=0.4) the differences are even more subtle.

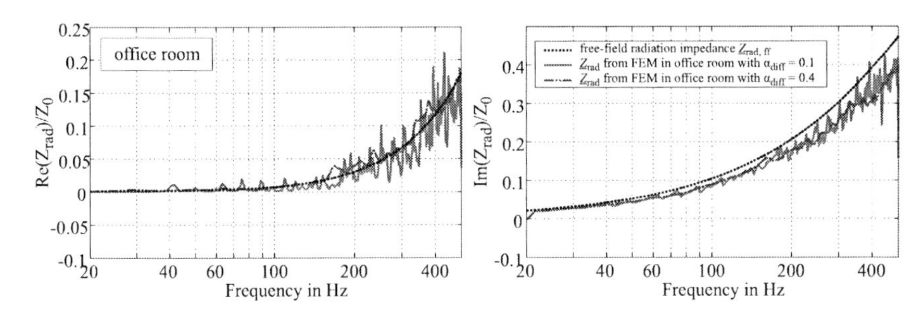

Figure 7.22.: Comparison of radiation impedance $\underline{Z}_{\mathrm{rad}}$ of a piston source in an extended baffle in the case of free field radiation and in the case of radiation into an office room (two different absorption settings).

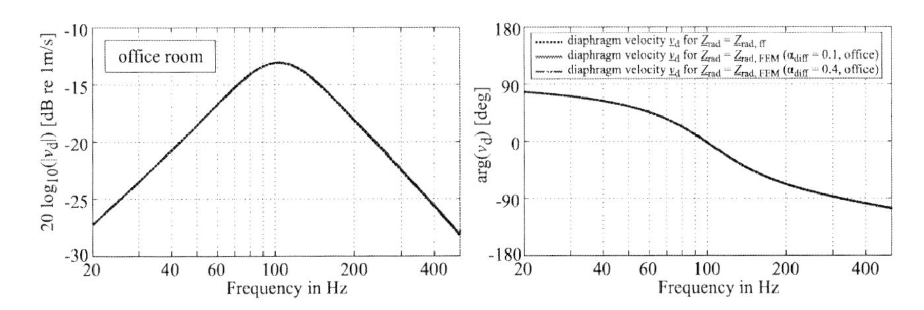

Figure 7.23.: Comparison of diaphragm velocity $\underline{v}_{\mathrm{d}}$ of a typical low- to midrange loudspeaker in the case of free field radiation and in the case of radiation into an office room (two different absorption settings).

With regard to room acoustic applications we can thus already conclude that despite the possible deviations between the actual and the free field radiation impedance (in the modally dominated frequency range of small, hardly damped rooms) the absolute value of the radiation impedance generally remains small compared to the sum of the other impedances and thus the influence of the room sound field on the diaphragm velocity is very small. Consequently, for room acoustic FEM simulations it is suitable to calculate the loudspeaker membrane velocity irrespective of the considered room using the Thiele-Small parameters and the free field radiation impedance (eq. 7.22).

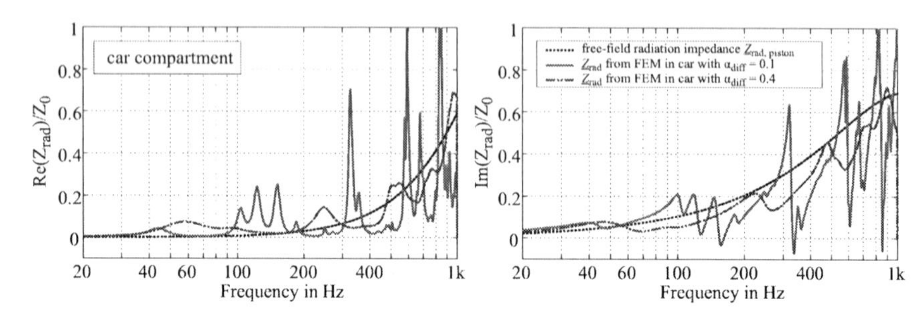

Figure 7.24.: Comparison of radiation impedance $\underline{Z}_{\mathrm{rad}}$ of a piston source in an extended baffle in the case of free field radiation and in the case of radiation into a car passenger compartment (two different absorption settings).

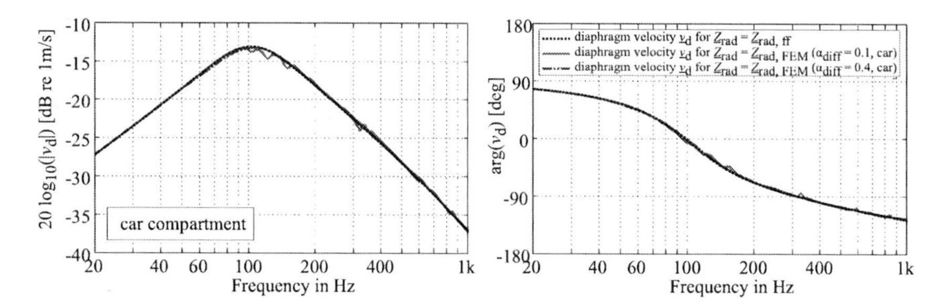

Figure 7.25.: Comparison of diaphragm velocity $\underline{v}_{\mathrm{d}}$ of a typical low- to midrange loudspeaker in the case of free field radiation and in the case of radiation into a car passenger compartment (two different absorption settings).

Finally, we will consider the sound field inside a loudspeaker box and its influence on the membrane velocity. In order to do so, we replace the lumped components representing the loudspeaker box ($C'_{\mathrm{ab}}, M'_{\mathrm{ab}}, R'_{\mathrm{ab}}$) by a complex impedance $\underline{Z}_{\mathrm{ab}}$ that is obtained from FEM simulations for three different types of damping in the box and set the acoustic radiation impedance for the front side of the loudspeaker diaphragm to $\underline{Z}_{\mathrm{ar1}} = \underline{Z}_{\mathrm{rad,ff}}(f)/S_{\mathrm{d}}$. Although this small investigation is a bit off-topic with regard to the room acoustic framework of this thesis, the results shall point to a possible application of a an electro-acoustically coupled FEM formulation. Figure 7.27 shows the influence of the differently damped closed loudspeaker box on the diaphragm velocity. This influence is in a first approximation described by an additional compliance $C'_{\mathrm{ab}} = \frac{V_0}{cZ_0}$ where V_0 is the cavity volume. It can be seen from the results in figure 7.26 that at very low frequencies ($< 100\,\mathrm{Hz}$) the back side radiation impedance of the loudspeaker diaphragm in the box agrees well with the simple lumped spring approximation. However at higher frequencies this back side radiation impedance is strongly dependent on the modal structure inside the loudspeaker box which again depends on the damping inside the enclosure. Depending on the type of the used loudspeaker and box these cavity modes can also affect the diaphragm velocity, so that in some cases an application of the FEM method for the determination of the cavity radiation impedance might be preferable to a simple lumped spring approximation. A possible application of the fully coupled electro-acoustic FEM model as given in the appendix might therefore be the investigation of the sound field inside a damped multi-way loudspeaker box.

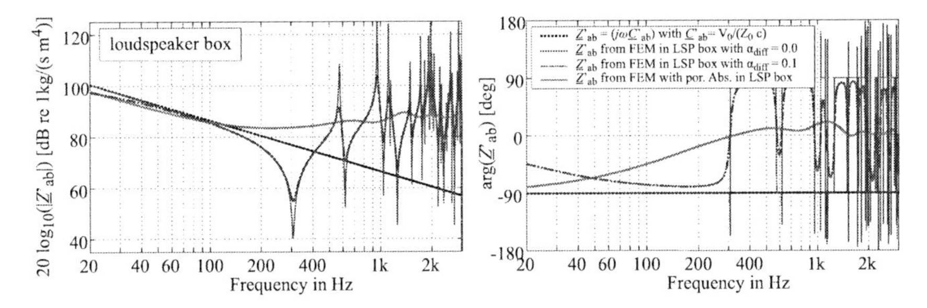

Figure 7.26.: Back side radiation impedance $\underline{Z}_{\text{box}}$ of a moving piston in a rectangular loudspeaker box for differently damped cavities compared to compliance model.

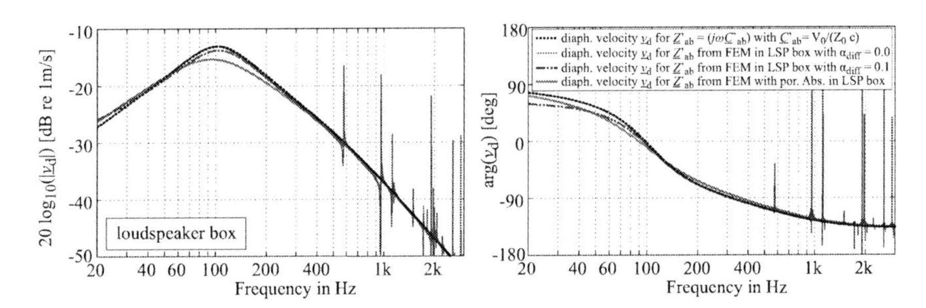

Figure 7.27.: Diaphragm velocity $\underline{v}_d$ of a typical low- to midrange loudspeaker in a closed box. The back side radiation impedance $\underline{Z}_{\text{box}}$ of the loudspeaker box is alternatively modeled by a lumped spring or by fluid FEM calculations for differently damped cavities.

8. Combined FE-GA simulations in a recording studio control room[1]

Based on the framework for the realistic simulation of sound fields in small rooms that was presented in the previous chapters, the next two chapters investigate the challenges and limitations of the combined FE-GA simulation approach on the basis of the comparison of room acoustic simulations and measurements in real, existing rooms with complex geometries, boundary and source characteristics. In particular, this chapter presents the measurement and simulation results for an existing recording studio control room and the following chapter 9 presents the according results obtained for a car passenger compartment. In both studies special emphasis is put on the best-possible determination of all necessary boundary and source characteristics which are obtained using the methods described in sections 5 and 6. Whenever possible, a variety of methods was used to determine the data for each considered boundary material/sound source, so that an assessment of the applicability of the different methods can be conducted on the basis of the obtained results. Moreover, the comparison of the measured and simulated results is based both on a direct comparison of the obtained room transfer functions/impulse responses as well as on a psycho-acoustically motivated comparison of the time and frequency dependent loudness characteristics of the results.

The results in chapter 7.2 have shown that in the case of a well defined room geometry and boundary conditions a very good agreement between measured and simulated low frequency sound fields can be achieved using the FE method. In the present section we now apply the combined FE-GA simulation approach to a more complicated setup, i.e. a recording studio control room, and benchmark the quality of the simulation by a comparison with measured monaural room transfer functions obtained in the real room. Special attention in the discussion of the results is paid to the low frequency FE prediction, which is crucial for the design of high quality studio acoustics. Furthermore, we describe all relevant factors that influence the prediction quality of the room acoustic simulations in the FE and GA domain with a focus on the specification of realistic boundary and source conditions for the simulation model.

[1]Early results of the presented study regarding the FE simulations in the considered recording studio were already published in Aretz and Maier [2008] and Aretz [2009]. Further results of the ongoing investigations and the efforts to improve the simulation boundary data and to extend the simulations to the whole audible frequency range were presented in Aretz et al. [2010b] and Aretz and Maier [2011]. The present section gives a conclusive up-to-date summary of the results of the whole study including completely new simulation results with an extended FE frequency range and improved GA algorithms. Moreover the present section presents so far unpublished aspects including (a) an investigation of the uncertainty in the room acoustic measurement results, (b) an extensive comparison of material boundary conditions obtained with different measurement methods and (c) first subjective evaluations of the comparison of simulated and measured room transfer functions. The subsections describing the recording studio, the boundary materials and the simulation setups are in parts taken from the above mentioned publications.

In the following a short outline of the study is given. First, sections 8.1 and 8.2 present the considered recording studio control room and the corresponding room acoustic simulation model used in the FE and GA domain. Next, section 8.3 describes the room acoustic measurements in the real room and presents the corresponding results. Sections 8.4 - 8.6 then give a detailed discussion on the determination of the boundary and source data that was used for the simulations, where special emphasis is put on the modeling of the various absorber types used in the room. Finally, section 8.7 compares the measured and simulated room transfer functions and section 8.8 presents the results of a preliminary subjective evaluation. To conclude, section 8.9 gives possible explanations for the observed discrepancies.

8.1. Real room

The considered recording studio is the private mermaid music studio in Munich, Germany, which was planned and realized by 'HMP Architekten und Ingenieure GbR / concept-A' who are also based in Munich. The floor plan of the recording studio including the considered source and receiver positions is shown in Figure 8.1. As can be seen in this figure the studio is built as a room-in-room construction using mineral wool backed plaster board panels which are mounted on a stud frame in front of the thick solid concrete walls. In order to achieve appropriate room acoustics in the control room various broadband and resonance absorbers are mounted to the side walls and the ceiling of the room. Moreover, special Helmholtz resonator boxes in the room corners have been designed and tuned for damping of the lowest eigenfrequencies in the room. Figure 8.2 shows photographs of the control room from different perspectives. A detailed discussion of the used sound absorbers, which significantly influence the sound field in the room will be given in section 8.4.

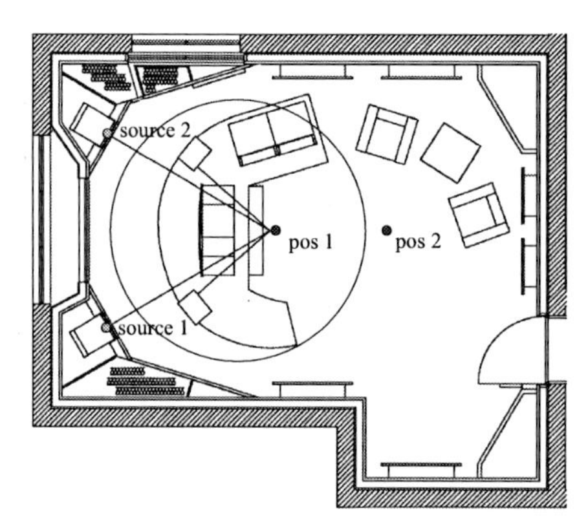

Figure 8.1.: Floor plan of the recording studio control room with considered source and receiver positions at 1.2m height

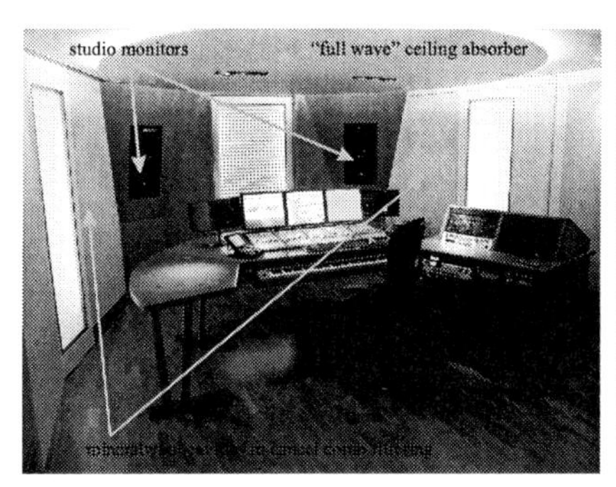

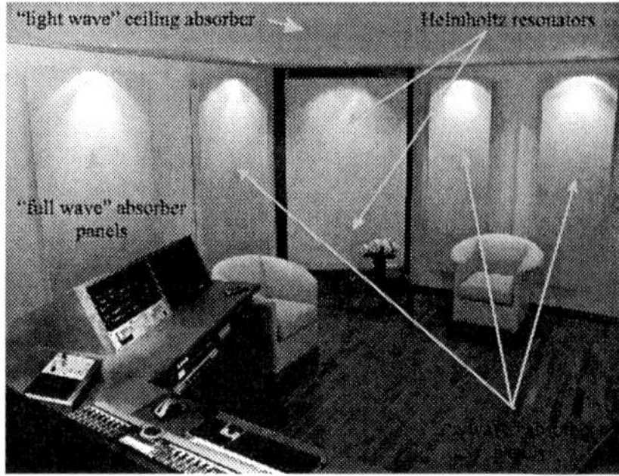

Figure 8.2.: Photographs of the control room at mermaid music (front and back view)

8.2. Simulation model

On the basis of detailed architectural plans of the control room a CAD model of the room geometry with all acoustically relevant design features was built. Based on this CAD model a finite element mesh and a triangulated surface model were created for the FE and GA simulation respectively. Figure 8.3 shows the FE mesh and the GA model of the control room (without the ceiling). All relevant acoustic features are marked in different colours and explained in the legend next to the figure. Additionally, the figure gives some information on the geometrical and room acoustic key parameters of the studio room.

Despite the rather low Schroeder frequency in the room at around 130 Hz the cross-fade frequency between FE and GA simulations is set to 500 Hz in order to account for the fact that the measured RTFs in the room still show pronounced dips and peaks up to at least 300 Hz (cf. figure 8.4). In order to generate a sufficient overlap for the filter cross-fade at 500 Hz, the upper frequency limit of the FE results was set to 800 Hz. The combination of FE and GA results was conducted following the guidelines given in section 4.3. Based on the constraint given in equation 4.9 the frequency step width of the FE simulations was gradually increased from $1 - 3$ Hz over the simulated frequency range, which appears sufficiently fine considering that the measured reverberation time decreases monotonically from 0.4 s to roughly 0.2 s within the considered FE frequency range. In particular a frequency step width of 1 Hz between $5 - 80$ Hz, 2 Hz between $82 - 500$ Hz and 3 Hz between $503 - 800$ Hz was used.

The geometrical acoustics results on the other hand are calculated for an image source order of three, a number of one million ray tracing particles and a detection sphere radius of 10 cm. Despite our results from section 7.1, which imply that in special cases the ISM can be used for the realistic prediction of low frequency sound fields, it is emphasized that even with a much higher image source order and the consideration of complex frequency dependent reflection factors a better prediction of the modal structure of the low frequency sound field is not expected in the considered room. This is mainly due to the fact, that the dimensions of neighboring regions with very different material conditions are not much larger than the considered wavelength up to 500 Hz. If we consider for example the back

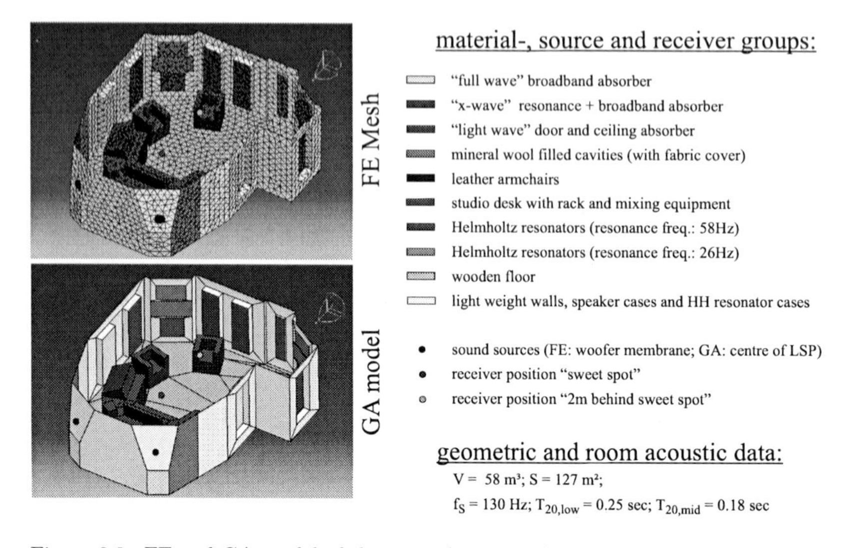

Figure 8.3.: FE and GA model of the control room with acoustically relevant objects. The given reverberation times $T_{20,\mathrm{low}}$ and $T_{20,\mathrm{mid}}$ were calculated as the average results obtained with a dodecahedron source in 3 different positions in the room, where the subscript 'low' indicates the average of the 63, 125 and 250 Hz octave band and the subscript 'mid' the average of the 500 and 1000 Hz octave band. The Schroeder frequency is then calculated using the average low frequency reverberation time.

wall of the room, we find that in the IS domain some image sources would be hardly damped because they hit the hard and reflective plasterboard walls and others would be heavily damped because they hit the absorber panels. This is however not a suitable model for the sound reflection of low frequency sound waves on such a wall.

8.3. Measurements in the real room

In order to compare our simulations to measured data, impulse response and reverberation time (RT) measurements were carried out in two independent measurement sessions in 2007 and 2010 in the real recording studio control room. In both measurement sessions it was tried to reproduce the simulation scenario shown in figure 8.3 to the best possible extend. We therefore removed all relevant objects from the room that were not considered in the simulation model, moved the leather arm chairs in the approximate position used in the simulations and carefully positioned the measurement microphone in the intended positions. By comparing photographs taken of the room in both measurement sessions we tried to make sure that the measurement situation was reasonably reproduced despite the long time distance between the measurements. The measurements of the first session have been conducted using Monkey Forest measurement software[2]. In the second session the

[2]Detailed information on the measurement software *Monkey Forest* can be found on the webpage http://www.four-audio.com/en/products/monkey-forest.html (last viewed: Jan. 2012)

ITA Toolbox for *Matlab*[3] was used for the playback, acquisition and post-processing of the data. For all measurements exponentially swept sine signals were used, from which the room impulse responses were deduced by deconvolution. As a measurement microphone we used a free field equalized 1/2" *Brüel & Kjær* microphone in all measurements.

The room impulse responses for comparison with the simulated RIRs were measured consecutively from the left and right studio monitor to the sweet spot position at the sound engineer's desk and a position two meters behind the sweet spot (as indicated in figure 8.1). A comparison of the low frequency RTFs measured in 2007 and 2010 from the left studio monitor to receiver position 1 is shown in the upper left part of figure 8.4. The results show a very similar low frequency response despite the long time distance and the inevitable small changes in the room setup between the measurement sessions. While a similarly good agreement in the low frequency response was found for the measurements using the right studio monitor and receiver position 1 (not reported here), it has to be admitted that the low frequency transfer function from the left studio monitor to position 2 (shown in the upper right part of figure 8.4 (b)) shows considerable deviations already in the octave bands below 200 Hz (with similar results found for the right studio monitor and receiver position 2). Considering the good match obtained for position 1 it is not entirely clear where these strong deviations stem from. Possible reasons might be attributed to an inaccurate positioning of the measurement microphone or possibly to an influence of the experimenter in the room who ran the measurements. Despite the fact that we tried to reproduce the measurement scenario with all due care in each case, the low frequency responses appear to be sensitive to some difficult to determine influencing factors. Taking further into account that we find a very good match between the FE simulations and the 2010 measurements for the left loudspeaker and the 2nd receiver position (see section 8.7) we come to the conclusion that the 2007 measurements for receiver position 2 were influenced by some unwanted and undisclosed factor.

Due to the observed stochastic fluctuations in the mid and high frequency bands of the measured room transfer functions it is not reasonable to focus on the fine structure of the impulse response in these frequency bands. For frequencies higher than 1 kHz we have therefore averaged the room transfer function in figure 8.4 in 1/6 octave bands. It can be seen that the energy in the mid and high frequency bands is very similar in both measurement sessions. The slightly higher levels obtained in the 2007 measurement for frequencies above 4 kHz may be attributed to a changed sensitivity curve of the *Brüel & Kjær* microphone, due to a different microphone orientation in both measurement sessions. In the 2007 measurements the microphone was pointing to the center of the stereo basis of the two monitor loudspeakers and in the 2010 measurements the microphone was pointing to the ceiling. Taking into account that the sound field in the studio is strongly dominated by the direct sound, the different orientations may account for the increased high frequency levels in the 2007 measurement. The difference in the sensitivity curve for different microphone orientations can be checked in the data chart of the *Brüel & Kjær* microphone.

[3]Detailed information on this acoustic signal processing and data acquisition toolbox, which is currently developed at ITA of RWTH Aachen University, can be found on the webpage `http://www.ita-toolbox.org/` (last viewed: Jan. 2012)

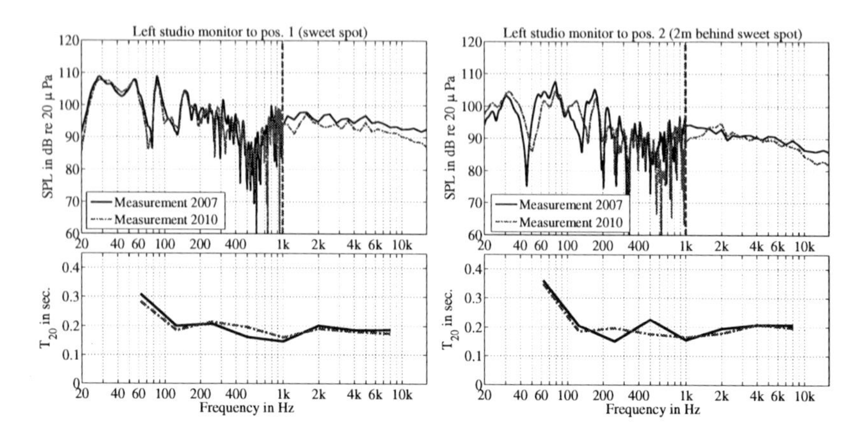

Figure 8.4.: Comparison of RTFs and RTs measured at two independent sessions in the recording studio control room.
The studio monitor in the left room corner (source1) was used as the source and the two receiver positions were located at the sweet spot position (pos1) and a a position 2 m behind the sweet spot (pos2), both at a height of 1.2 m. The results were obtained at two independent measurement sessions in 2007 and 2010. The room transfer functions are averaged in 1/6 octave bands for frequencies higher than 1 kHz.

Additionally, for both measurement sessions reverberation times were calculated from the sweep measurements with the left studio monitor loudspeaker mounted in the room corner to receiver position 1 and 2. The bottom part of figure 8.4 shows the results of these measurements with an acceptable agreement of the reverberation times obtained in both cases.

It is important to note that the observed order of magnitude of differences between the independent measurement results gives important implications regarding the demanded accuracy of the room acoustic simulations in the considered room.

8.4. Determination of absorption and impedance data

Under the simplifying assumption that all room boundaries in the recording studio can be considered as locally reacting surfaces, which is generally a good approximation for porous materials and hard reflective surfaces, their acoustic behaviour can be modeled by applying appropriate frequency dependant impedance boundary conditions in the FE model. Since it is possible to deduce the absorption coefficient from the acoustic surface impedance but not unambiguously vice-versa, consistent material data for the FE and GA simulation can most easily be obtained from broadband impedance data. However, since the reliable measurement of the acoustic surface impedance for the whole audible frequency range turns out to be very difficult with commonly used measurement setups like the impedance tube or the Microflown in-situ setup (see section 5.1 and 5.2), the 1D-equivalent network model described in section 5.3 was used to calculate the acoustic surface impedances of the main absorbers in the recording studio. The third octave band absorption coefficients for the GA simulation were then calculated from the impedance data to get consistent data in both models.

Instead of using normal incidence data, both absorption coefficients and acoustic surface impedances were calculated for field incidence conditions using equations 3.11 and 7.16 with an upper integration limit of 78°. While using diffuse field or field incidence absorption coefficients is state-of-the-art in GA, the use of field incidence impedances is justified based on the findings in section 7.2. At this point, it should be mentioned again that for locally reacting boundary materials the normal incidence impedance fully captures the angle dependency of the absorption characteristics, which is in this case solely caused by the change in the characteristic impedance of air $Z_0' = \frac{Z_0}{\cos\theta}$ (cf. section 3.2.2 for more details). Thus for a locally reacting material $\underline{Z}_{\text{field inc.}} = \underline{Z}_{\text{S}}(0)$ but $\alpha_{\text{field inc.}} \neq \alpha(0)$.

The following subsections give a detailed description of the different boundary materials in the room and discuss how the acoustic surface impedances and absorption coefficients have been determined for the simulations.

8.4.1. Layered broadband and resonance absorbers

In the control room three different types of layered broadband and resonance absorber configurations are used, which are denoted 'fullwave', 'x-wave' and 'lightwave' according to the naming convention of the manufacturer 'concept-A acoustics' (see figures 8.2 and 8.3 for more information on the positioning of the absorbers in the room.). The absorbers are composed of different layers of porous materials, airtight foils, air gaps and fabric covers. In order to calculate the field incidence impedance and absorption characteristics of the three absorber types the single layers of the two-port network model have been parameterized using detailed construction and material data which was kindly supplied by the 'HMP' architects. The resulting field incidence data which was used in the simulations in the FE and GA simulations of the studio is shown in figure 8.5. Since the exact composition of the absorbers is proprietary information of 'HMP' Architects it can however not be described in this thesis.

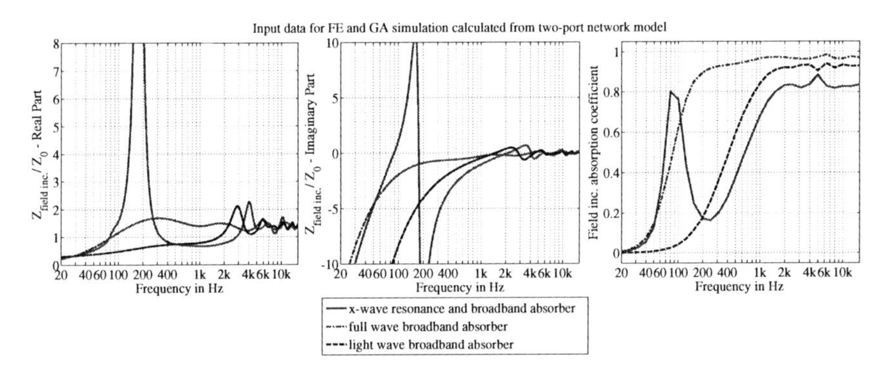

Figure 8.5.: Specific surface impedance and absorption data of the layered studio absorbers as used in the FE and GA simulations of the recording studio.
The data was calculated for field incidence conditions (0° – 78°) using the two-port network model for layered absorbers.

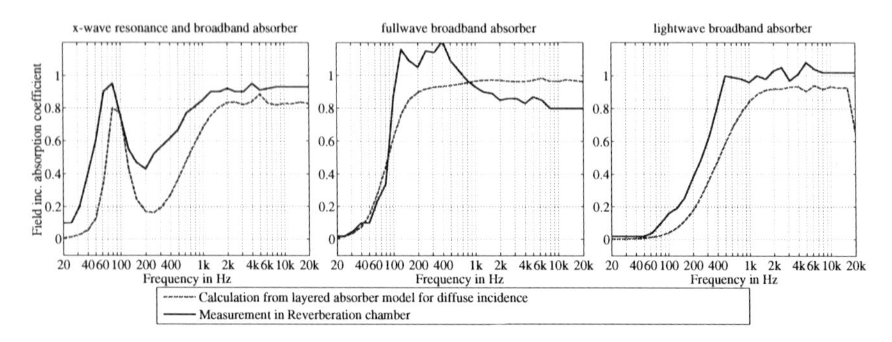

Figure 8.6.: Comparison of field incidence absorption coefficients of the studio absorbers calculated from the two-port network model and measured in a reverberation chamber.

In addition to the two-port network model calculations, extensive impedance and absorption measurements were conducted on the 'fullwave', 'x-wave' and 'lightwave' absorbers. In particular, measurements were carried out in the ITA impedance tube and with the Microflown method both under in-situ conditions (in the recording studio) and under laboratory conditions (in the anechoic chamber at ITA of RWTH Aachen University). Additional measured absorption results obtained in a reverberation chamber were supplied by the 'HMP' architects.

Figure 8.6 shows the calculated third octave band field incidence absorption values in comparison to the measured reverberation chamber data. It can be seen that the reverberation chamber results mostly exceed the predicted results from the two-port model and for some frequencies even absorption coefficients greater than 1 were measured. This is believed to be due to considerable edge effects at the boundaries of the absorbers. On the other hand the reverberation chamber measurements clearly show the predicted low frequency resonance behaviour of the 'x-wave' absorber, and an overall acceptable agreement is found regarding the frequency ranges and slopes for the transition from the low to the high absorption regime.

In order to compare our impedance tube and Microflown measurements to the predictions of the two-port network model, figure 8.7 shows the specific surface impedances and absorption coefficients of the layered absorbers for normal sound incidence. The measurements in the impedance tube and with the Microflown in-situ setup were averaged over a number of measurements as indicated in the legends of the plots. Despite the overall acceptable agreement between the results obtained with the different methods, the following observations can be made:

- Although the comparison of the impedance data obtained with all different available methods reveals a considerable measurement uncertainty of this parameter, the agreement of the absorption characteristics is generally good.

- Overall the Microflown measurements in the anechoic chamber show the best match with the two-port model data with an analyzable frequency range from < 100 Hz up to 20 kHz. It is however important to mention that the given analyzable frequency range of the laboratory in-situ measurements with the Microflown setup can surely not be generalized to arbitrary absorber types. Instead it depends both on the absorber size, shape and absorption characteristics. Detailed information on the factors influencing the results obtained with the in-situ Microflown setup are given in the Master's thesis by van Gemmeren [2011]. Although the measurements in the anechoic chamber were carried out on samples of exactly the same size as those used in the studio room ($1.75 \times 0.75\,\mathrm{m}^2$), the in-situ measurements of the absorbers installed in the studio room only yield reasonable results above roughly 250 Hz. This is due to the fact that a very short time-window had to be applied to cancel strong, early reflections from surrounding objects. In contrast to this, the anechoic measurements could be processed with a wider and frequency dependent time windowing, which made it even possible to measure the low frequency resonance behaviour of the 'x-wave' absorber (below 100 Hz).
- As already mentioned in section 5.1 the measurements in the cylindrical ITA impedance tube can only be analyzed up to a frequency limit of 8 kHz. While the impedance tube measurements of the 'fullwave' and 'lightwave' absorber yield high quality absorption coefficients down to the 60 Hz third octave band, the low frequency resonance behaviour of the 'x-wave' absorber (below 100 Hz) could not be measured correctly, which is due to clamping effects of a 1 mm thick and heavy airtight foil in the absorber configuration. This can be understood by considering that in order to only work as a surface mass with $\underline{Z}_{\mathrm{layer}} = j\omega m'$, it would be necessary to eliminate all edge effects and thus mount the foil frictionless but still airtight in the impedance tube. Despite the use of a vast amount of vaseline and careful cutting of the foil samples this appeared however impossible considering the small diameter of the tube. The use of a larger impedance tube would possibly alleviate this problem.

It should be mentioned for the sake of completeness that it would in principle also be possible to include coupled 3D models of each absorber layer in the FE domain. This would however necessitate appropriate models for porous materials and thin membranes in the FE domain and a very fine discretization would be needed to simulate the wave propagation in the thin absorber layers. In addition to the considerably increased meshing complexity and computation time, another major drawback from a practitioners point of view is that such an approach would necessitate different room models in the FE and GA domain. It is therefore emphasized that the presented approach, which uses (a) consistent field incidence impedance and absorption data and (b) the same room model in the FE and GA domain, offers the low frequency FE extension at an only marginal additional cost with regard to the simulation setup compared to a pure geometrical acoustics simulation. Moreover, it was concluded from section 7.2 that the used field incidence impedance model constitutes an adequate physical model of the sound reflection at locally and laterally reflecting porous boundaries. Taking further into account the inevitable uncertainty in the determination of the material parameters for each absorber layer, we believe that for room acoustic applications the possible small benefit of a full 3D modeling of the absorbing boundaries is generally outweighed by the immense additional cost of these calculations.

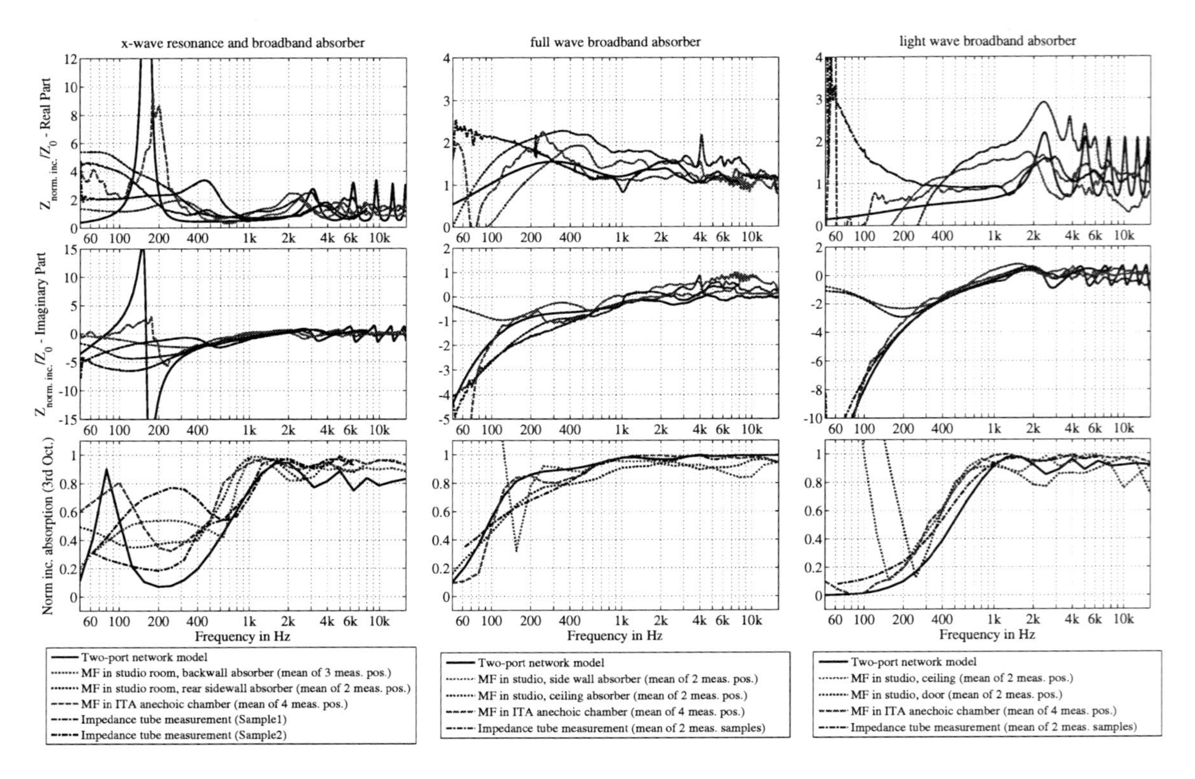

Figure 8.7.: Comparison of normal incidence impedances and absorption coefficients of the studio absorbers calculated from the two-port network model and measured in the impedance tube and with the Microflown in-situ method.

8.4.2. Helmholtz resonators

The Helmholtz resonator boxes are installed in the rear corners of the control room and are hidden behind big rectangular textile screens (cf. figures 8.2 and 8.3). The resonance frequencies of the Helmholtz resonators are tuned to the two lowest eigenfrequencies of the room which lie at approximately 26 and 58 Hz. In addition to the large textile screens a second fabric is fixed directly in front of the resonator holes in order to further reduce the Q-factor of the resonator system.

With regard to the FE modeling of the Helmholtz resonators we can apply a similar reasoning as above concerning the benefits and drawbacks of a full 3D model versus an impedance boundary condition. While at first sight the former appears to be the more accurate model, it has two major drawbacks. Firstly, an inclusion of the full resonator geometry (i.e. the fluid domains in the resonator holes and the cavity) into the FE mesh would again necessitate the generation of a more complex and different room model as used in the GA simulation and secondly, the consideration of a fabric cover in front of the resonator holes (as commonly used) would require the inclusion of an appropriate coupled FE model, which models the sound transmission through a thin flow-resistive layer with fluid domains on both sides. Such a model is described by Aretz [2008a]

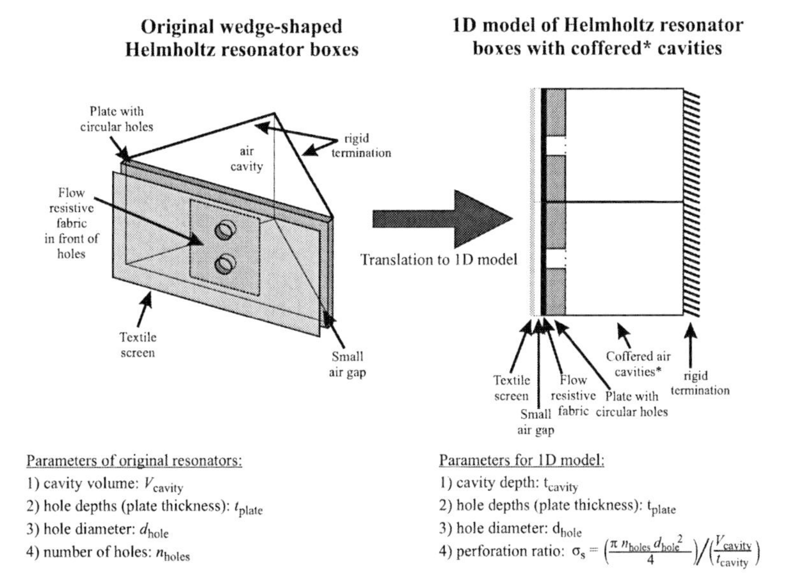

Figure 8.8.: Modeling of the Helmholtz resonator boxes in the recording studio using the two-port network model for layered absorbers.
* "coffered" means in this context that even for angular incidence the sound propagation within the resonator boxes is restricted to the direction perpendicular to the front plate. This is a reasonable assumption for the low frequency range in which the considered resonators are active and it is important when calculating the field incidence impedance and absorption characteristics of the Helmholtz resonator.

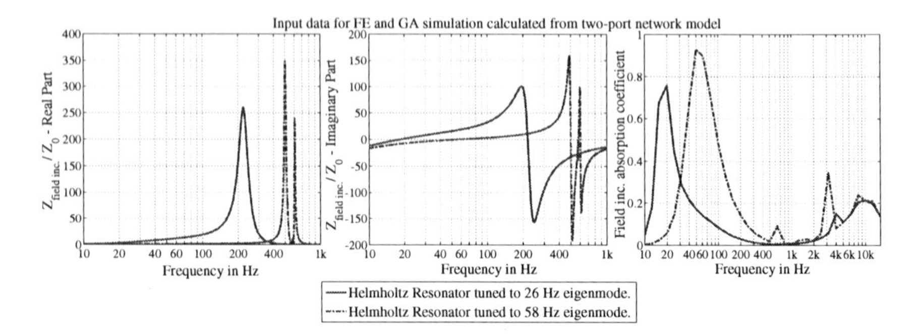

Figure 8.9.: Specific surface impedance and absorption data of the Helmholtz resonator boxes as used in the FE and GA simulations of the recording studio.
The data was calculated for field incidence conditions ($0° - 78°$) using the two-port network model for layered absorbers.

who derives an FE formulation which couples the two-port network approach for locally reacting layered absorbers (cf. section 5.3) to fluid domains on both sides of the absorber configuration. A summary of this FE-model is also given in the appendix of this thesis. However, with regard to Helmholtz resonator boxes it has been shown by Aretz [2008b] that a damped Helmholtz resonator box can be adequately modeled by a surface averaged acoustic impedance in room acoustic FE simulations.

In the course of this study the damped Helmholtz resonator boxes were therefore modeled by a surface averaged impedance boundary condition based on the two-port network model described in section 5.3. A detailed description of the layer configuration consisting of the fabric covers with a small air-gap in-between, the perforated plate and the air cavity as well as the transition of the real 3D resonator box to the corresponding 1D model is given in figure 8.8. The geometric parameterization of the perforated plate and cavity layers was conducted on the basis of detailed plans of the resonator boxes supplied by the 'HMP' architects. The resulting field incidence surface impedance and absorption coefficient data which was used in the FE and GA simulations is given in figure 8.9. It can be seen that the used two-port network model adequately models the resonance behaviour of the damped resonator boxes and that the absorption characteristic of the 58 Hz resonator box well matches the resonance frequency specification given by the 'HMP' architects. However, the Helmholtz resonator for the 26 Hz eigenmode shows a slightly lower resonance frequency as specified by the 'HMP' architects. This deviation turns out to be due to the flow resistivity in front of the resonator holes. Calculations of the impedance characteristics of the undamped resonator yield a resonance frequency of almost exactly 26 Hz. However, the flow resistivity in front of the holes shifts this resonance frequency to approximately 20 Hz. It should finally be mentioned that it was not possible to verify if the damping of the resonance curves matches the damping of the real Helmholtz resonators well. This can only be guessed on the basis of the simulation results for the entire room which will be discussed in section 8.7.

8.4.3. Lightweight wall construction

The lightweight construction in the recording studio consists of three screwed layers of 12.5 mm thick plasterboard which are mounted at 100 mm distance to a 240 mm thick solid concrete wall. For damping reasons the 100 mm gap is filled with mineral wool. According to the findings presented in Franck and Aretz [2007] such a construction can efficiently be modeled in the FE domain using an impedance boundary approach, which models the lightweight construction as a damped mass-spring system. This is firstly due to the fact that the mineral wool layer can in good approximation be considered as locally reacting. Secondly, considerable influence of the plate eigenmodes or coincidence effects on the sound field in the room are only expected at the very low end of the FE frequency range[4]. Moreover a realistic simulation of the plasterboard with its eigenmodes using a 2D thin shell model would necessitate knowledge and consideration of the exact mounting and clamping conditions of the plates, which is not known in the considered case.

On the basis of the above made reasoning the field incidence surface impedance of the lightweight construction is therefore modeled by a surface mass backed with a porous mineral wool layer in front of a rigid termination. It should be mentioned that formulas exist to correct the surface mass term m' for coincidence effects. The effective surface mass is in this case given as a function of the angle of incidence θ_0 [Mechel, 1998]:

$$\underline{Z}_\mathrm{S} = j\omega m'_\mathrm{eff}, \text{ with } m'_\mathrm{eff} = m'\left[(1-(\frac{f}{f_\mathrm{cr}})^2\sin^4\theta_0) - j(\eta(\frac{f}{f_\mathrm{cr}})^2\sin^4\theta_0)\right], \tag{8.1}$$

where f_cr is the critical frequency of the considered elastic plate. According to Mechel [1995, p.772] these formulas are however only valid up to frequencies which lie at least half an octave below the critical frequency f_cr. The mass correction was therefore only applied in the calculation of the field incidence surface impedance for the FE frequency range. In contrast to that the absorption characteristics for the GA frequency domain were calculated using a simple surface mass $j\omega m'$ which yields an almost completely rigid behaviour of the lightweight walls above 100 Hz. Due to the high surface mass of the three screwed layers of plasterboard, the error which is introduced by the negligence of coincidence effects in the GA domain is considered to be small. The resulting field incidence surface impedance and absorption coefficient data which was finally used in the FE and GA simulations is shown in figure 8.10.

8.4.4. Mineral wool filled cavities

As indicated in figures 8.1 and 8.3 the large cavities between the Helmholtz resonator boxes and those next to the big wooden loudspeaker cases in the room corners are filled up with mineral wool which is hidden behind a textile cover. These areas provide high absorption from the highest to the lower-mid frequency bands. In particular, the purpose of

[4] With a density of $\rho = 880\,\mathrm{kg/m^3}$, a thickness of $d = 37.5\,\mathrm{mm}$, lateral dimensions of $2.3 \times 1.0\,\mathrm{m^2}$ and a reasonable estimate for the Young's Modulus and Poisson ratio of $E = 3.2\,\mathrm{GPa}$ and $\nu = 0.3$ [Mechel, 1998, p.309] the plaster board plates have a critical frequency of $f_\mathrm{cr} \approx 900\,\mathrm{Hz}$ and the lowest plate eigenmode at $f_0 \approx 40\,\mathrm{Hz}$ (under the assumption that all edges are simply supported). Between these frequencies the plate behaviour is dominated by its mass effect, which is accounted for in the impedance boundary approach.

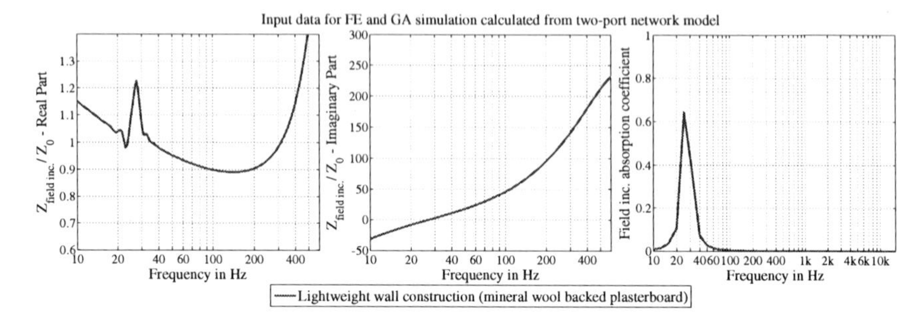

Figure 8.10.: Specific surface impedance and absorption data of the damped lightweight wall construction as used in the FE and GA simulations of the recording studio.
The data was calculated for field incidence conditions ($0° - 78°$) using the two-port network model for layered absorbers. The specific impedance was calculated using the surface mass correction as given in equation 8.1 for the plasterboard plates, whereas in the calculation of the absorption data a simple surface mass $j\omega m'$ was used.

the absorbing areas next to the loudspeaker cases is to cancel the first lateral reflection of the respective other stereo loudspeaker, in order to avoid comb filtering effects. However, since the density or flow resistivity of the used mineral wool could not be determined in the course of this study, it appears difficult to predict the low frequency absorption characteristics of these areas. The impedance and absorption characteristics therefore needed to be estimated using the two-port network model, where the absorber was modeled as a single layer porous absorber with moderate flow resistivity of $8\,{}^{\mathrm{kPas}}/_{\mathrm{m^2}}$ and a thickness of 300 mm for the cavities between the Helmholtz resonator boxes and 800 mm for those next to the big wooden loudspeaker cases. This roughly corresponds to the depth of the cavities, which was determined from figure 8.1.

8.4.5. Other surfaces

The remaining surfaces in the room were mainly considered as almost rigid with a small estimated broadband absorption. These surfaces comprise the wooden floor, the wooden loudspeaker cases in the front corners of the room, the sides of the absorber cases and the wooden boxes for the Helmholtz resonators, which were all given an estimated broadband absorption of 0.075. In order to account for the diverse equipment on the studio desk during the measurements the desk was given an estimated broadband absorption of 0.15. The absorption of the leather coated armchairs was estimated to 0.3. Better absorption values for the armchairs could of course be derived from absorption measurements in a reverberation chamber. Unfortunately, this was not feasible in the course of this study.

The total absorption of these surfaces was admittedly chosen with an eye on the reverberation time in the room. However, since (a) the estimated values lie within a reasonable range for the considered surfaces and (b) the total absorption of the unknown surfaces is comparatively small compared to the main absorbers in the room, it is believed that the chosen estimations have no strong detrimental influence on the quality of the simulation results.

8.5. Determination of scattering data

The scattering coefficients needed for the GA simulation of the studio room have not been measured in the course of this study. However, most of the surfaces in the room are flat and thus scattering effects in the higher frequency ranges appear to be mostly negligible in the room. On the other hand, it was already discussed that scattering/diffraction effects presumably play an important role in the low frequency part of the room transfer function due to the diffraction effects at the edges of the absorber panels. However, since on the one hand the low frequency part of the RTF is covered by the FE results and on the other hand the determination of realistic low frequency scattering coefficients for all surfaces appears to be extremely difficult, it was not tried to model or estimate this behaviour in the GA simulations. Therefore frequency constant scattering estimates of 0.1 were applied to most of the surfaces in the room. Strong high frequency scattering effects were only expected on the studio desk, which had lots of recording equipment on it during the measurement. Its scattering coefficient was therefore set to a broadband value of 0.5. While the chosen values admittedly only constitute very rough estimates, we can report that GA simulations with slightly changed scattering coefficients (between 0.0 and 0.3) did not lead to noticeable variations in the simulation results.

8.6. Source and receiver characterization

As can be seen in figure 8.2 the studio monitors used in the measurements and simulations are mounted flush into special wooden cabinets in the room corners such that the sweet spot of the stereo basis is located at the engineer's seat behind the studio desk. In order to account for the characteristics of the studio monitors in the GA simulation it is necessary to account for their directivity pattern and pressure transfer function. In the present study measured free field pressure and directivity data was used, which was kindly supplied by IFAA[5] and which was obtained using the swivel arm measurement setup shown in figure 6.1 (b). Since during the IFAA measurements the loudspeaker box was not mounted flush into a large flat panel as in the studio, slight differences between the measured and the actual directivity and pressure TF data have to be expected due to diffraction effects at the boundaries of the loudspeaker box. The directivity data on the rear hemisphere of the loudspeaker was accordingly set to zero for the simulations. Figure 8.11 shows the horizontal and vertical directivity maps of the used loudspeaker for the frontal hemisphere.

[5]Institut für Akustik und Audiotechnik (http://www.ifaa-akustik.de).

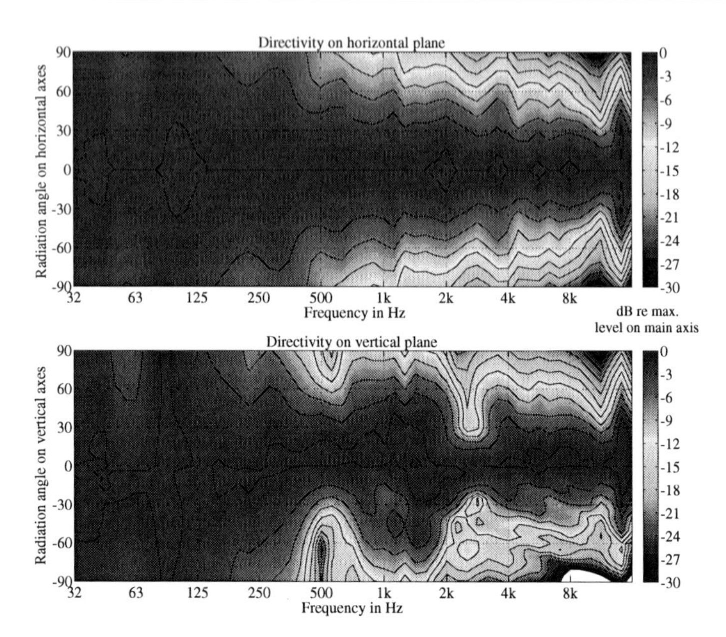

Figure 8.11.: Horizontal and vertical directivity maps of the used K&H O500 loudspeaker.

In the FE simulation the source characteristics are accounted for by the membrane velocities and, since the used studio monitor is a bass reflex enclosure, also the velocity in the ports. Since the cut-off frequency of the loudspeaker woofer is at 520 Hz, mostly the woofer and the bass reflex holes contribute to the sound field in the considered FE frequency range (< 500 Hz). As already discussed in section 6.1.1, the direct measurement of the diaphragm and port velocities would however require laborious additional measurements (with e.g. a laser vibrometer and/or a Microflown velocity sensor). In the present study the approach presented in section 6.1.1 (cf. equation 6.2) was therefore used to calculate an equivalent piston velocity from the measured free field pressure transfer function. This velocity was then assigned to a piston surface placed in the position of the woofer membrane in the FE model (cf. figure 8.3). The pressure contributions of the membranes (mostly the woofer in the FE range) and the ports are thus summed up in a single piston source which would generate the same on-axis transfer function at 1 m distance in the free field if mounted flush into an infinite baffle. Figure 8.12 shows the measured free field pressure TF at 1 m distance and the calculated velocity function, which are both given for 1 V at the loudspeaker input. Possible issues in the FE simulation might be due to a different directivity pattern of the idealized piston source compared to the real multi-way loudspeaker source. These problems will be discussed in the results sections.

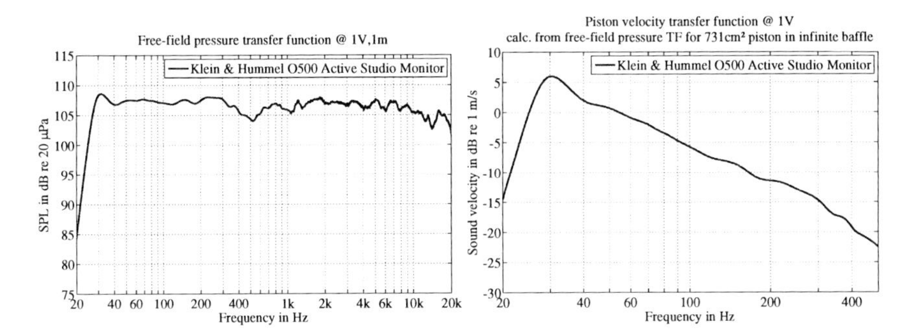

Figure 8.12.: On axis free field pressure sensitivity at 1 V, 1 m of used K&H O500 loudspeaker and the deduced piston velocity transfer function for the FE calculations.

8.7. Comparison of simulation and measurement results

Figure 8.13 shows a comparison of the measured and simulated low frequency RTFs, the band averaged energy levels and the reverberation characteristics in the recording studio obtained for receiver position 1 and 2[6]. It can be seen in the upper part of the figure that the FEM simulations reproduce the strong modal structure of the sound field below 300 Hz well. Above this frequency however, the FE simulation fails to predict the fine structure of the transfer function in all its detail. This is believed to be due to the fact, that at these frequencies the modal overlap is already quite high (considering the Schroeder frequency at approximately 130 Hz) and thus the transfer function is already very sensitive to slight errors in the prediction of the single room modes. This notion is also supported by the fact, that the measured RTFs obtained in the two independent sessions in 2007 and 2010 also show noticeable deviations in this frequency range.

For frequencies higher than 400 Hz the simplified piston source model in the FE domain may also contribute to the overall error between measured and simulated (FE) RTFs, since the interference effects of the real loudspeaker generated at the crossfade between the woofer and midrange driver are not considered in the FE source model. As can be seen in figure 8.11 the crossfade between the woofer and midrange driver generates a considerable distortion of the directivity of the loudspeaker on the vertical axis. This is not considered in the idealized piston model used in the FE simulation. Moreover, above 500 Hz it is mostly the much smaller midrange driver that radiates the sound from the

[6]Taking into account the observed differences in the measurement results obtained in 2007 and 2010, the measurement results, which are plotted in this section and which were also used for the auralizations, were obtained by crossfading the 2007 and 2010 measurements at 350 Hz in order to get a best possible fit with the simulated data. In particular we have taken the low frequency part of the 2010 measurements (since the simulated low frequency part of the RTF in receiver position 2 shows a considerably better match with these results) and the mid- and high frequency part of the 2007 measurements (since the overall level of the GA simulations better corresponds to the high frequency levels of the 2007 measurements).

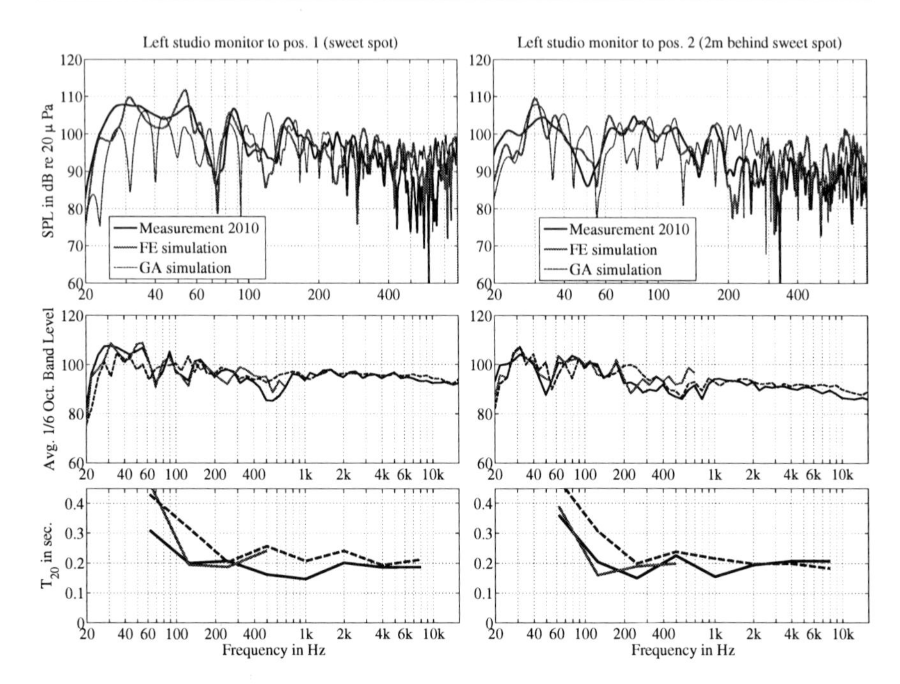

Figure 8.13.: Comparison of measured and simulated room transfer functions and reverberation times from the left studio monitor to two receiver positions in the recording studio control room.

studio monitor. By comparing the measured directivity of the studio monitor with that of the idealized piston source in the frequency range from 500 to 800 Hz, it was found that the measured directivity of the studio monitor shows a noticeably stronger focussing of the sound to the principal direction than the idealized source. Taking further into account that the FE piston source velocity was calculated based on the measured on-axis free field pressure transfer function, it can be concluded that the idealized FE piston source radiates more energy in these frequency bands than the real loudspeaker source. This reasoning corresponds well with the noticeably overpredicted FE sound pressure levels between 400 and 800 Hz, although it is obviously difficult to interpret all details in the observed differences between measured and simulated RTFs. However, despite the observed differences between the low frequency RTF obtained from the measurement and the FE simulations, the plot clearly shows the considerable improvement which is obtained by using the FE low frequency transfer function instead of the pure geometrical acoustics based simulation.

The middle part of figure 8.13 shows the band averaged energies of the measured and simulated RTFs. Thanks to the inherent normalization of the absolute level of the FE and GA simulations (see section 4.3) the results blend in very well at the cross-fade frequency without any further adjustment. Moreover, it can be seen that in the range

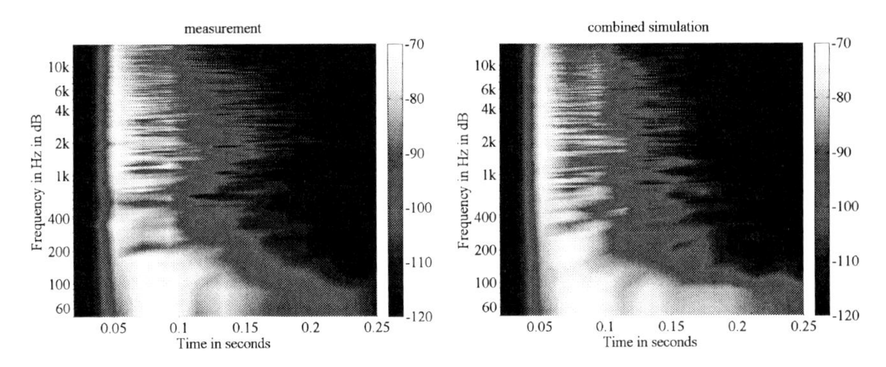

Figure 8.14.: Comparison of spectrograms generated from measured and simulated room transfer functions from the left studio monitor to the sweet spot receiver position in the recording studio control room.

above the transition frequency between the FE and GA results the band energy is generally very well predicted by the GA simulations. Finally, the lower part of figure 8.13 shows a comparison of the measured and simulated reverberation times. Taking into account that the simulations are fully based on the impedance and absorption data obtained from the two port network model and the rough broadband estimates for the unknown low absorbing surfaces, the reverberation characteristics of measured and simulated room transfer functions show an already good match. It is clear that a perfect match between measured and simulated reverberation times can easily be achieved by iteratively adjusting the impedance and absorption data. However, this is not within the scope of the present study. Instead we want assess the best possible quality of room acoustic simulations that are exclusively (or at least almost exclusively) based on a-priori determined input data.

A more detailed picture of the time and frequency dependency of the measured and simulated room impulse responses is given in figure 8.14 which shows the spectrograms obtained in the sweet spot position with the left studio monitor as the source. Despite an observed slight underestimation of the damping in the lowest frequency bands (which can equally be deduced from the higher quality factors of the lowest room modes in the simulated RTF and also from the reverberation times) the spectrograms corroborate the overall good match between measured and simulated data.

8.8. Preliminary subjective assessment of results

A preliminary subjective comparison of the simulated and measured room transfer functions was conducted by convolving the measured and simulated monaural impulse responses obtained for the two measurement positions with different audio material ranging from speech to classical, pop and hip hop music. Due to the very short reverberation time in the studio it did not appear reasonable to directly compare the impulse responses. The resulting audio files were then played back via headphones to a small number of trained listeners in order to assess possible differences in a direct A/B comparison of the corresponding files. Moreover, the convolved audio files were compared with regard to their time and frequency dependent loudness characteristics, in order to investigate if the subjective assessment of the listeners corresponded to the differences in the loudness curves.

The aim of the subjective comparison of measured and simulated audio files was on the one hand to assess the absolute quality of the combined simulation approach and on the other hand to answer the question, if the objectively improved simulation accuracy in the low to lower-mid frequency range (obtained with the FE method) is also appreciated in a subjective evaluation when compared to a broadband geometrical acoustic simulation. The measurements were therefore not only compared to the combined simulation results but also to the full bandwidth geometrical acoustics results. The calculations of the specific and total loudness levels were conducted following the guidelines given in 'DIN 45631/A1: Berechnung der Lautheit zeitvarianter Geräusche' with an update rate of 2 ms and no time overlap. For the loudness calculations the convolved audiofiles where multiplied by a constant factor, such that the RMS level of the measurement based auralizations was equal to 84 dB. It should be emphasized that the same multiplication factor was used for the auralizations based on measured and simulated IRs. Thus, the level differences between the measured and the simulated IRs was kept exactly the same as seen in figure 8.13.

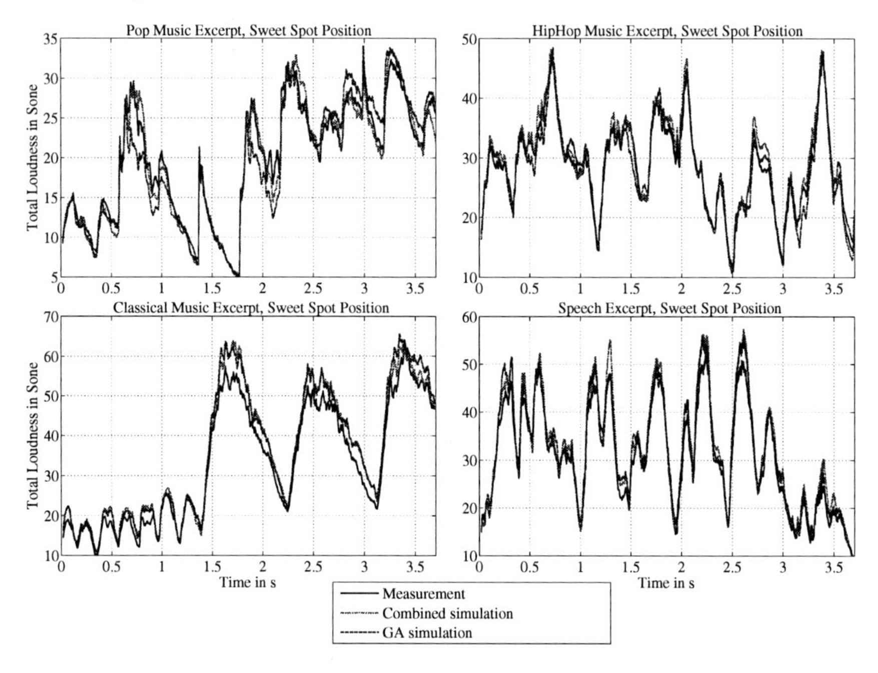

Figure 8.15.: Total loudness of different audio excerpts convolved with measured and simulated room impulse responses in the recording studio.

The total loudness curves for the four different sound stimuli convolved with the measured and simulated room impulse responses are given in figure 8.15. Moreover, figures 8.16-8.19 show the differences between the specific loudness over time for the first nine critical bands. Finally, table 8.1 shows the time averaged percentage difference of the specific loudness values obtained from the auralization files with the measured and simulated impulse responses. Details regarding the calculation of the percentage difference are given in the annotations of the table.

The results in table 8.1 indicate an overall benefit of the FE extension at least in the critical bands up to 300 Hz. However, by looking at the more detailed specific and total loudness plots in figures 8.16-8.19, it becomes clear that this result is not unambiguous. Although the plots for the pop and hip hop stimuli indeed support the implications from table 8.1 by showing an improved match of the low frequency loudness characteristics calculated from the combined simulation and measurement results (especially in the two lowest critical bands), this result is unfortunately not apparent in the loudness plots of the classical music and the speech stimuli. With regard to the total loudness curves these stimuli show hardly any differences between the combined and the pure GA simulation, which can partly be attributed to the fact, that the total loudness of these stimuli is mainly driven by mid frequency components which lie above the cross-fade frequency of the FE simulation. However, at least in the case of the classical music excerpt, it appears that the specific loudness of the combined simulation shows an even slightly worse match with the measured results than the pure GA simulation. When trying to compare the results in the total and specific loudness plots with the condensed results in table 8.1 it is thus important to account for the fact, that the time instances and bark bands with only very small contributions to the total loudness equally contribute to the time averaged percentage difference as those with high contributions to the total loudness. Thus the table gives an average assessment of the relative difference in the first eight critical bands while the total and specific loudness plots show the absolute difference.

The results from the evaluation of the loudness characteristics were mostly confirmed by the judgments of the listeners, who argued that improvements related to the FE extension were if at all noticed in the pop music and hip hop music samples which contained strong bass frequency components generated by either a bass drum or a synthesizer bass. In contrast to this, the combined simulation based auralizations of the classical music and the speech were not judged significantly better than the pure geometrical acoustics results. Overall the simulation quality was rated as very high and none of the listeners reported any artefacts from the combination of the FE and the geometrical acoustics results. However, all listeners were able to recognize differences between the stimuli based on the measured and simulated impulse responses, although none of them was able to guess which stimuli was based on the measured IR and which was based on the simulated IRs (both combined and GA-only).

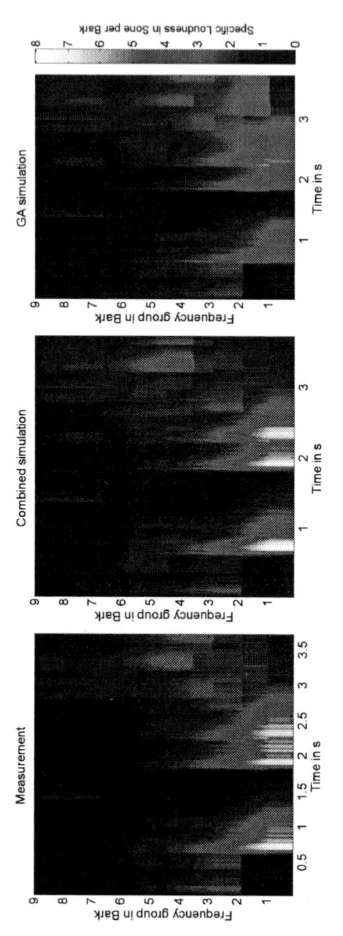

Figure 8.16.: Specific loudness of pop music excerpt convolved with measured and simulated room impulse responses.

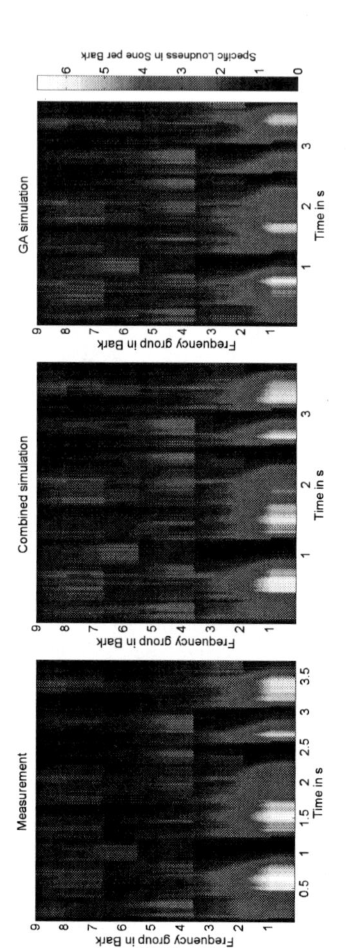

Figure 8.17.: Specific loudness of hip hop music excerpt convolved with measured and simulated room impulse responses.

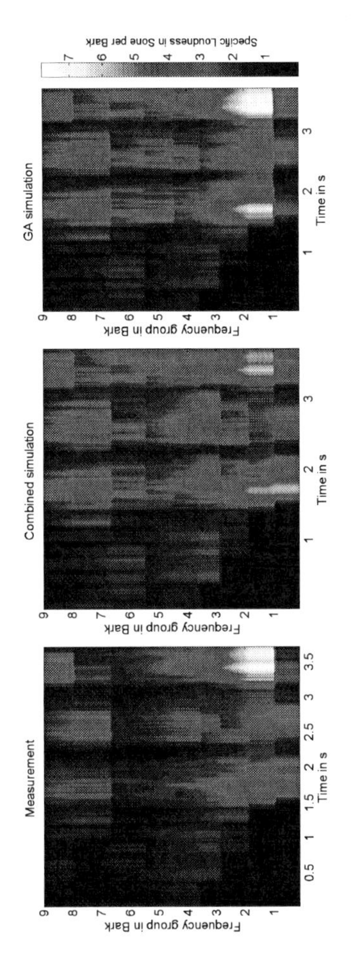

Figure 8.18.: Specific loudness of classical music excerpt convolved with measured and simulated room impulse responses.

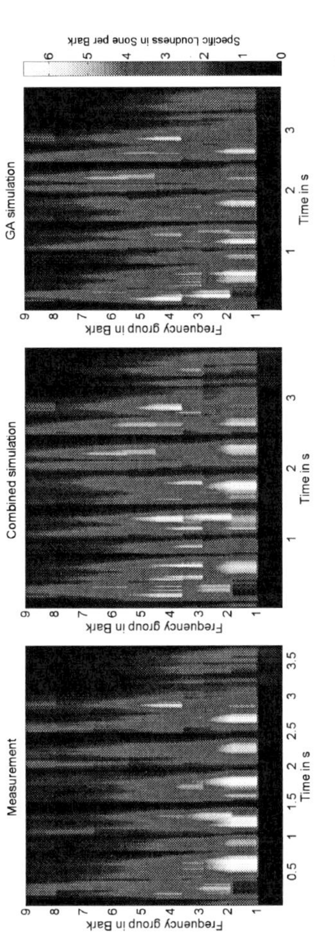

Figure 8.19.: Specific loudness of speech excerpt convolved with measured and simulated room impulse responses.

Table 8.1.: Time averaged percentage difference between meas. and sim. based auralization files.

Bark Band	Freq.-range [Hz]		Pos1: Sweet Spot Pop	HipHop	Classics	Speech	Pos2: 2 m behind sweet spot Pop	HipHop	Classics	Speech	MEAN
1	20-100	*Comb. Sim.*	11.63	9.14	18.05	11.20	5.84	25.68	15.93	7.98	**13.18**
		GA Sim.	20.54	20.21	31.02	19.63	15.67	30.20	31.03	19.37	**23.46**
		Diff. (GA-Comb.)	8.91	11.07	12.96	8.43	9.84	4.52	15.11	11.39	**10.28**
2	100-200	*Comb. Sim.*	14.22	10.19	14.84	12.19	8.68	14.70	7.54	7.88	**11.28**
		GA Sim.	23.94	19.51	16.34	18.13	16.85	22.36	16.92	22.59	**19.58**
		Diff (GA-Comb)	9.72	9.31	1.50	5.94	8.17	7.66	9.38	14.72	**8.30**
3	200-300	*Comb. Sim.*	16.53	10.45	16.60	10.51	11.95	11.37	10.85	11.41	**12.46**
		GA Sim.	14.21	16.64	16.46	15.25	25.04	32.85	21.96	30.86	**21.66**
		Diff (GA-Comb)	-2.32	6.19	-0.14	4.74	13.09	21.47	11.11	19.45	**9.20**
4	300-400	*Comb. Sim.*	15.87	9.27	24.07	15.26	23.69	10.18	12.83	18.35	**16.19**
		GA Sim.	11.10	10.47	9.76	12.67	27.26	18.96	19.80	27.81	**17.23**
		Diff (GA-Comb)	-4.76	1.20	-14.31	-2.58	3.57	8.78	6.97	9.46	**1.04**
5	400-510	*Comb. Sim.*	16.98	14.72	20.66	22.47	27.32	15.65	15.27	23.40	**19.56**
		GA Sim.	13.03	13.76	16.15	16.41	24.13	15.31	17.99	22.30	**17.39**
		Diff (GA-Comb)	-3.95	-0.96	-4.50	-6.07	-3.19	-0.33	2.73	-1.10	**-2.17**
6	510-630	*Comb. Sim.*	30.51	26.66	40.58	32.40	25.89	17.51	24.37	27.88	**28.23**
		GA Sim.	23.47	25.12	42.14	24.19	20.89	13.50	15.82	22.03	**23.39**
		Diff (GA-Comb)	-7.04	-1.54	1.56	-8.21	-5.01	-4.01	-8.55	-5.85	**-4.83**
7	630-770	*Comb. Sim.*	27.04	27.97	48.44	34.71	18.77	14.90	17.59	29.31	**27.34**
		GA Sim.	22.10	27.34	48.52	26.63	15.91	12.86	14.52	24.19	**24.01**
		Diff (GA-Comb)	-4.94	-0.63	0.08	-8.08	-2.86	-2.04	-3.07	-5.12	**-3.33**
8	770-920	*Comb. Sim.*	18.44	16.28	38.72	23.58	12.26	12.79	13.60	22.25	**19.74**
		GA Sim.	17.93	16.19	38.51	17.96	12.10	12.33	13.01	17.98	**18.25**
		Diff (GA-Comb)	-0.51	-0.09	-0.21	-5.61	-0.16	-0.47	-0.59	-4.27	**-1.49**

The table shows the time averaged percentage difference between the monaural auralizations based on the measured and simulated room impulse responses. The time averaged percentage difference is calculated both for the combined as well as for the pure geometrical acoustics simulation for four different sound stimuli and two different receiver positions. The last column in the table gives the mean value over all stimuli and both receiver positions.

Since the specific loudness is calculated for a critical band rate (tonality) resolution of 0.1 Bark and a time update rate of $t = 2$ ms (as suggested in 'DIN 45631/A1'), the mean specific loudness within each critical band is obtained by first averaging all ten values within each band. The time averaged percentage difference is then calculated from the specific loudness values $N'(z,m)$ for each critical band as

$$d(z) = \frac{1}{M}\sum_{m=1}^{M} \frac{|N'_{\text{meas}}(z,m) - N'_{\text{sim}}(z,m)|}{0.5 \cdot (N'_{\text{meas}}(z,m) + N'_{\text{sim}}(z,m))}, \quad \text{with} \quad z = 1, 2, \ldots, 24 \tag{8.2}$$

where z is the critical band number and m is the time index for the 2 ms time steps. The table also reports the difference $d_{\text{GA}}(z) - d_{\text{Combined}}(z)$, where a positive value indicates that the combined results are on average closer to the measurements and vice versa for the negative values.

8.9. Discussion and summary

Despite the overall good agreement between measured and simulated frequency responses there naturally remains room for improvement or at least for discussion, since it will be explained that some theoretically possible improvements may not be exploitable due to limitations in the simulation methods or in the determination methods for the boundary and source conditions. Moreover, it should be mentioned that a similar discussion regarding possible improvements and open questions was also given in our first publication on FE simulations in the considered recording studio [Aretz, 2009]. To our satisfaction, most of the previously raised points have been addressed in the course of the present study and we have found that some measures did indeed help to improve the simulation results, while others did not or at least not noticeably. The following discussion therefore gives an up-to-date summary of possible or impossible improvements regarding the key factors influencing the simulation quality.

1. **Uncertainties due to the simulation methods**

   The finite element part of the simulation covers all relevant wave effects of the sound propagation in the recording studio. Thus the simulation quality is mainly driven by the quality of the geometric room model and the source and material data. However, it has to be mentioned that even for an exact geometric model and ideal source conditions, the impedance boundary approach is limited by the inherent assumption of locally reacting boundaries. On the other hand, it has been discussed at length in section 8.4 that this assumption is in most cases admissible and that even for non-locally reacting materials the introduced error may be alleviated by using field incident average impedances (cf. section 7.2).

   In contrast to the FE method the GA method uses many simplifications regarding the sound propagation and sound reflection in the room. While it was found that the time and frequency distribution of the mid and high frequency energy in the impulse response is nonetheless well captured by the GA simulations (cf. figure 8.14), possible problems in the fine structure of the reflection pattern of the impulse response can be attributed to the negligence of diffraction effects and the difficulty of determining realistic low frequency scattering coefficients for the used materials. It is thus emphasized, that in rooms with rather small adjacent material regions with very different boundary conditions a physically motivated lower frequency bound for the applicability of the GA simulation presumably lies well above the Schroeder frequency.

2. **Uncertainties due to the room model**

   As described above the room model was build from detailed architectural plans of the recording studio room and features all acoustically relevant objects in the room. Although the model neglects some small objects and geometric details in the room, we believe that a further refinement of the room model would generally not lead to better simulation results, since (a) these geometric details do not considerably affect the modal structure of the low frequency sound field and (b) the acoustic effects of these details at higher frequencies are more adequately and more efficiently modeled by assigning reasonable scattering coefficients to the respective objects in the geometrical acoustics domain. Special attention in this context should be paid to the studio desk with the computer screens and diverse equipment on it, since a rather strong first order reflection is contributed by these surfaces (at least to the receiver position in the sweet spot directly at the desk). According to the studio builders the reflection/diffraction at the studio desk is even a possible reason for the mid-frequency dip (between $500-1000\,\text{Hz}$) in the transfer function at the sweet spot position, which is neither well captured in the FE nor the GA simulation. While this reasoning is surely a bit speculative it can in any case be asserted that a realistic consideration of the diffraction and scattering effects on the table appears hardly possible in the GA domain. On the other hand, in the FE domain improvements could, if possible at all, only be achieved by a further refined geometric model and boundary data of the table with the screens and equipment on it.

With regard to the topic of geometric detail in the simulation models it should be mentioned that a current project at the ITA of RWTH Aachen University explicitly deals with the influence of geometry simplifications in the geometrical acoustic domain and first interesting results were recently presented by Pelzer and Vorländer [2010] and Pelzer et al. [2010b]. Although these studies deal with much larger rooms, they generally support our notion that excessive geometrical detail does not necessarily improve the subjectively perceived quality of geometrical acoustics simulations.

3. **Uncertainties due to the boundary conditions**

   With regard to the simulation input data it has to be emphasized that except for the source representation the simulation results are purely based on the boundary data obtained from the theoretical two-port network model and that no measured absorption or impedance data was used for the simulations. The high quality of the obtained simulation results thus confirms the applicability of the two-port network model for the determination of realistic boundary data for room acoustic simulations. In contrast to the measured boundary data as obtained in the impedance tube or with the Microflown setup, the two-port network model has the big advantage that it yields plausible and consistent acoustic surface impedance and absorption data for the whole audible frequency range (and for arbitrary angles of incidence) for various types of absorbers. Moreover, while measured data for the 'x-wave', 'fullwave' and 'lightwave' absorbers (obtained in the impedance tube or with the Microflown setup) yielded on average similar reflection characteristics compared to the two-port network approach, the measured data showed considerable restrictions with regard to the usable frequency range and in some frequency bands also systematical errors due to unavoidable deficiencies of the measurement methods (e.g. clamping effects in the impedance tube or diffraction effects caused by finite sample sizes for the Microflown setup). It is therefore emphasized that despite the considerable effort and care in the measurement of the boundary data, no actual benefit could be gained from the measured impedance results compared to the results obtained from the two-port network approach. Additionally, it has to be taken into account that both the used materials and the fabrication process underlie a certain variability and thus a determination of the reflection characteristics with the two-port network model based on representative material data supplied by the manufacturer appears very reasonable and saves valuable time, compared to laborious impedance measurements using the above mentioned methods.

As was already mentioned in section 8.4 the use of full 3D models of the absorbers in the room does also not appear efficient in the considered case. This is on the one hand due to the high additional modeling complexity and the increased computation times required for this approach and on the other hand due to the considerable difficulties and inaccuracies that go along with the measurement and evaluation of all relevant material parameters and mounting conditions.

Finally, it should be mentioned that the absorption coefficients that were used for the studio desk and the arm chairs are of course only very rough estimates and therefore present another source for inaccuracies in the simulation. While an estimate of the average, frequency dependent absorption characteristics of these objects can in principle be obtained in a reverberation chamber measurement, the determination of a reasonable complex acoustic surface impedance of these materials appears to be hardly feasible with the available measurement methods.

4. **Uncertainties due to the source representation**

   Regarding our sound source representation two possible sources for errors can be identified, which may noticeably deteriorate the simulation results; (a) in the FE domain we have combined the contributions of the bass reflex holes and of the membrane velocities in a single equivalent piston source; (b) the used free field loudspeaker data (used directly in the GA simulations and to calculate the equivalent piston velocity in the FE simulations) was obtained for the loudspeaker box mounted on a swivel arm and not as in the studio mounted flush into an extended flat panel. While it is unfortunately not possible to exactly quantify in how far these factors influence the simulated room transfer function, we can report from our experience that regarding the free field room transfer function no pronounced narrow-band differences are expected between the monitor loudspeaker box being mounted flush into an extended baffle or not. Moreover, since for both receiver positions the very strong direct sound contribution comes almost from the principal direction of radiation of the loudspeaker, strong effects due to small errors in the directivity pattern for lateral radiation are also not expected. Finally, it can be stated that the combination of the source contributions of the bass reflex holes and membrane velocities, appears surely unproblematic in the frequency bands, where mostly the woofer contributes to the sound generation. However, in and above the crossfade range between woofer and the midrange driver at about 520 Hz, the simplified piston source model may introduce some errors, due to the wrong directivity pattern compared to the real loudspeaker box (cf. section 8.7).

5. **Summary of subjective evaluation**

   With regard to the preliminary subjective evaluation it is also possible to draw some interesting conclusions regarding the overall simulation quality and the benefit of the low frequency FE extension. Especially the discussions with the listeners as well as the evaluation of the time and frequency dependent loudness characteristics gave insightful implications for future work.

a) Without a direct comparison to the measurement results, both the combined and the broadband GA simulation sounded equally plausible and realistic to the listeners. In particular, none of the listener's reported any artefacts or audible deficiencies in the simulations, which indicates that (a) the used combination method for FE and GA simulations does not produce any audible artefacts and (b) the inadequate representation of the modally dominated low frequency sound field below the Schroeder frequency in the pure GA simulations does not lead to any conspicuous anomalies in the auralizations.

b) Despite the considerable effort in the determination of best possible simulation input data and the overall high simulation quality (with regard to the low frequency RTF, the band energy levels and the reverberation time), all listeners were able to hear a difference between the auralizations based on measured and simulated (combined and pure GA) impulse responses. These differences were mostly attributed to a slightly different coloration of the sound stimuli.

c) Despite the fact that the objective comparison of the measured and simulated low frequency transfer functions shows an obvious improvement of the combined simulation approach over the pure GA simulations, a subjective improvement is only perceived for stimuli with high energy contributions in the bass frequency range and even then only by careful listening.

d) The results of the comparison of the low frequency RTFs as well as those of the loudness evaluation indicate that the benefits of the low frequency FE extension clearly lie in the modally dominated part of the sound field. In particular, we were not able to show any benefit of the FE simulations in the frequency range above approximately 300 Hz, which is due to the already high modal overlap in this frequency range. At least in the considered room (without large diffracting objects) we therefore do not expect any further improvement of the overall simulation quality by pushing the upper frequency limit of the FE calculations even higher. At least not without further refinement of the boundary and source data as well as the room model, which is, as discussed above, very difficult to achieve.

9. Combined FE-GA simulations in a car passenger compartment[1]

The present section summarizes the results of a three-year research project on the simulation of sound fields in car passenger compartments for the whole audible frequency range using the combined FE-GA approach outlined in section 4. The project was carried out in collaboration with a partner in the automotive industry and aimed at the generation of realistic binaural auralizations of the sound field generated by the loudspeakers of the car audio system inside the passenger compartment. The general challenges faced in the project were of course very similar to those described in the simulation study of the recording studio presented in the previous section. However, as will be shown in the following, the car passenger compartment constitutes a far more difficult simulation environment than the recording studio control room. This has various reasons. Firstly, a car passenger compartment is obviously not designed with the primary intent of exceptionally good acoustics (in contrast to a recording studio). While this is in itself no problem for the simulation, it conveys the problem that relatively few or none a-priori information exists on the acoustic characteristics of the boundary materials and sound sources in the car. Even worse, in contrast to the materials used in the studio (e.g. the homogeneously layered absorbers) the car materials are mostly not accessible to a straightforward determination of their acoustic characteristics, due to curved shapes and highly inhomogeneous material structures. Secondly, the geometrical acoustics simulation faces obvious limitations caused by (a) the various diffraction edges at the front seats and head rests, (b) the high geometrical detail and complex, curved shape of many boundary surfaces and (c) the close vicinity of sound source and receiver to each other as well as to surrounding boundaries (cf. limitations of the ISM and SRT in sections 4.2.1 and 4.2.2). Finally, the inclusion of a binaural receiver model in the FE and GA domain, also introduces open questions, since due to the very small volume of the car compartment it is a-priori not clear in how far the presence of one or several occupants influences the room acoustics in the car.

[1]Although intermediate results of the project have already been presented at international conferences in 2009 and 2010 [Aretz and Vorländer, 2009, Aretz and Knutzen, 2009, Aretz and Jauer, 2010] and more extensive project related results are reported in the master theses by Knutzen [2008] and Jauer [2010], it has to be emphasized that the present section gives a conclusive up-to-date summary of the project results including completely new simulation results. All simulations have been re-run for the following two main reasons; (a) the GA simulation tool RAVEN has undergone continuous, far reaching modifications and improvements in the course of the project and (b) our continuous efforts regarding the determination of better boundary and source data have lead to inevitable adjustments of the used simulation input data, which make it difficult to compare the results obtained at different stages throughout the project both in the FE and GA domain. The presented new results are thus obtained on the basis of consistent simulation input data using the latest RAVEN version (as of October 2011) and a thoroughly tested FE solver.

The present chapter investigates the above raised questions in order to find solutions and suggest guidelines for the resulting challenges in the simulations. All investigations in the course of the project are based on simulations and measurements of two different car models of the same manufacturer. While the sound sources, i.e. the loudspeakers in the car compartments, have been investigated individually for both car models, the material data has been obtained on the basis of laboratory measurements of sample materials and in-situ measurements which were only obtained for one of the car models. However, since the used materials in both car models are very similar, it appears reasonable to use the obtained boundary data also for the simulations of the other car model.

An outline of the chapter is given as follows. To start with, section 9.1 describes how the simulation models for the FE and GA simulations have been obtained from detailed CAD models of the car compartments which were supplied by our partner in the automotive industry. Next, sections 9.2 and 9.3 deal with the determination of all necessary boundary and source data, and discuss the difficulties, challenges and limitations in conjunction with the applied measurement methods. Based on this input data, section 9.4 presents selected room acoustic simulations for the two car passenger compartments and compares the results with the corresponding measurements obtained in the real compartments. While section 9.4 focuses on the objective comparison of the results, the following section 9.5 discusses the differences in the measured and simulated room transfer functions based on a preliminary subjective comparison of the results. Since the sound field in a real car passenger compartment is generally influenced by many variable influencing factors (e.g. occupancy in the car, adjustment of seat position, clothing of occupants) section 9.6 presents further measurement studies which aim at the quantification of the uncertainty in the room transfer functions caused by these factors. In order to conclude the study, section 9.7 discusses possible reasons for the observed differences between measured and simulated room transfer functions and summarizes the results.

It should finally be mentioned that for the sake of brevity we do not report all results that were obtained in the course the project. Instead, the focus is put on the presentation and discussion of selected, representative results that illustrate the potentials and limitations of the applied methodology in all steps of the simulation process including the determination of the required input data.

9.1. Simulation models

The used FE meshes and GA models were extracted from highly detailed CAD models of the considered passenger compartments of the two investigated car models. Figure 9.1 shows pictures of the original CAD geometries, which were supplied by our partner in the automotive industry, as well as the deduced FE meshes and GA models. While the FE meshes could be obtained with reasonable time and effort using commercially available, state-of-the-art meshing tools[2], similar tools for an automatic generation of suitable triangulated geometrical acoustics models with a reasonably small number of

[2]The meshes were generated with *CATIA* (`www.3ds.com/catia`) or *LMS Virtual Lab* (`http://www.lmsgermany.com/simulation/virtuallab`).

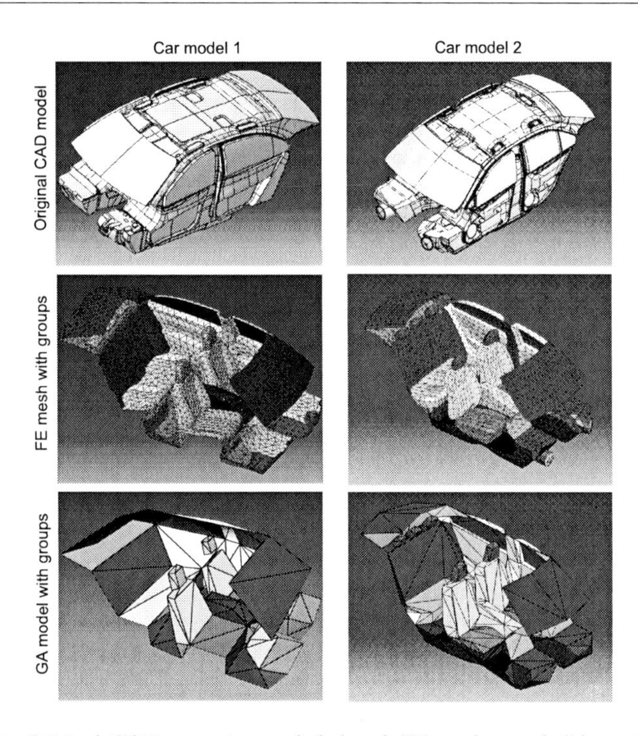

Figure 9.1.: Original CAD geometry and deduced FE meshes and GA models for car model 1 and 2.

faces and yet a sufficiently accurate representation of the car geometries could not be found. Instead, the triangulated GA models had to be generated by hand on the basis of the highly detailed car models of the car compartments. The models were generated using commercially available CAD tools[3] and finally exported for the use in *RAVEN*. It can be seen that both the FE meshes and the GA models capture all relevant geometric features of the respective car compartments. Moreover, as indicated by the different colouring, the model boundary is subdivided into disjunct domains in order to assign acoustic boundary conditions complying with the different materials used in the car compartments.

In order to generate true binaural FE results a dummy head model was optionally included in the FE mesh of one of the car models. Since in the binaural measurements a similar dummy head (without body) was used, we did refrain from including a full occupant's body in our FE model for better comparability with the measured results. A picture of the dummy head model, which was carefully trimmed and merged with the seat geometry to facilitate the meshing process, can be seen in figure 9.2. In the GA domain on the

[3]The triangulated GA models were generated using the CAD software *CINEMA 4D* (`http://www.maxon.net/de/home.html`) and the CAD Tools in *LMS Virtual Lab*.

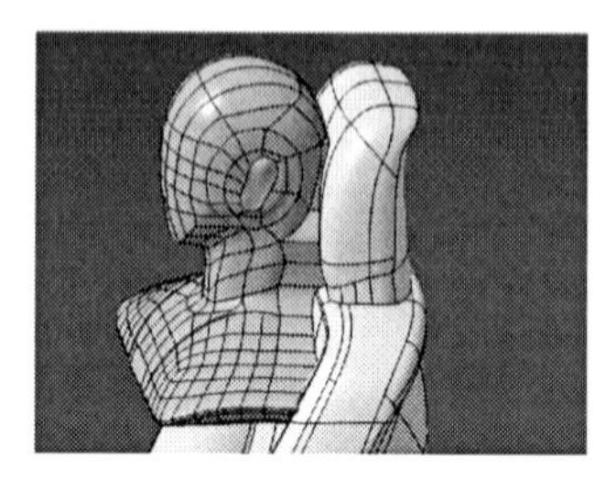

Figure 9.2.: Dummy head merged with car seat in CAD model of passenger compartment. The dummy head was considered in FE simulations to generate binaural room transfer functions.

other hand, the consideration of binaural cues is effectuated by the use of head related transfer functions (see section 4.2.4) and not by an inclusion of an actual head geometry. Thus the GA model did not have to be adapted for the consideration of binaural impulse responses. Possible issues regarding the receiver representation in the binaural FE and GA simulations are dealt with in section 9.4 which compares the measured and simulated binaural room transfer functions in the car compartment.

The used FE meshes (with and without dummy head) were meshed with parabolic tetrahedron elements using an average element size of approximately 80 cm, which allows for simulations up to at least 1.25 kHz. Additionally, an even finer mesh was generated for car model 1 with an average element size of 40 cm. This mesh was used to investigate if a further increase of the upper frequency limit of the FE simulation leads to improved simulation results. The frequency step width was set to 2 Hz, which easily meets the requirement of equation 4.9 considering the low measured reverberation times shown for example in figure 9.19. On the other hand, the GA models consist of 188 and 585 planar polygon surfaces respectively, which allows for a maximum image source reflection order of 2 using the current *RAVEN* implementation. Moreover, for the ray tracing part a number of 500000 particles, a time resolution for the energy histograms of 1 ms and a diameter of 14 cm for the receiver sphere was used in both models.

9.2. Boundary conditions

In order to specify realistic boundary conditions for the FE and GA model, extensive measurements of the car interior materials were conducted to determine their acoustic impedances and absorption coefficients using the methods described in section 5. Figures 9.3 and 9.4 summarize our efforts and gives a categorization and description of the different car materials with a list of the methods applied to each material to determine its acoustic reflection characteristics. The corresponding impedance and absorption results are given in the following subsections, which also discuss the uncertainty in the obtained results and the apparent problems and limitations when applying the different methods to the materials.

Car Material	*Picture*	*Material Characteristics*	*Impedance Tube*	*Microflown in-situ*	*Model Calculation*	*Reverberation Chamber*
seat cushion		Complex layered structure. Mostly porous layers, except for cover material (leather, fabric).	Good reproducibility if sample is fitted into the tube with care regarding exact size and compression of samples. Similar results for different areas of seat, although material layering is not homogeneous throughout the seat. Measurement with closed leather cover is difficult (leakage problems).	Measurements are blurred by interfering close reflections from other surfaces. Satisfactory results obtained with averaging. Similar results for different areas of seat cushion and backrest.	Characteristics of porous materials are generally well captured by model. However, considerable uncertainty in measurement of absorber parameters and problems due to inhomogeneity of material layers.	-
backrest		Complex layered structure. Mostly porous layers, except for cover material (leather, fabric).				-
car mats on carpet with foam backing		Car mats consist of porous carpet with stiff and airtight backing. The carpet underneath the car mats is also considered in the measurement.	Satisfactory results. However, problems due to inhomogeneous thickness of the heavy foam underneath the floor carpet. In the tube measurement the carpet was either directly backed by a rigid termination or was measured with an average thickness of the underlying heavy foam.	Additional absorber mats used to cancel close disturbing reflections. Satisfactory results for the mid- and high-frequency range	Porous layer and airtight backing of car mat and carpet could not be disassembled to measure single layer absorber parameters such as flow resistivity. Estimation of absorber layer parameters by fitting with impedance tube results	-
floor carpet with foam backing		The carpet in the foot space of the car is backed by a thin airtight sort of glue and a heavy foam of inhomogeneous thickness.				-
windows		Wind screen, rear window and side windows.	The impedance tube is not suitable for almost rigid material samples.	In-situ setup is not suitable. Very high uncertainty, because velocity is close to zero.	Adequate results. The windows are modelled as a surface mass with a free field termination.	-

Figure 9.3.: Categorization of car materials and summary of applied measurement methods (Part 1). A more detailed discussion is given in the respective subsections on the individual materials.

Car Material	*Picture*	*Material Characteristics*	*Impedance Tube*	*Microflown in-situ*	*Model Calculation*	*Reverberation Chamber*
headliner		Thin porous layer in front of dense backing mounted at approximately 5cm distance to the roof sheeting.	Unsatisfactory results. Air gap between headliner and roof sheeting is difficult to consider. Unwanted additional friction and spring stiffness caused by clamping issues on the tube walls.	Unsatisfactory results. Although the in-situ measurement considers the actual mounting of the headliner, the results are heavily blurred by diffraction effects and close interfering reflections from other surfaces.	Exact depth of air-gap is unknown. Porous layer and airtight backing could not be disassembled to measure single layer parameters. Absorber layer parameters could only be guessed by comparison with imp. tube and rev.chamber results	full headliner with roof sheeting was mounted in special rigid box and measured in the reverberation chamber. No reflection phase available.
door lining		The door lining and the dashboard are made of thin layers of wood or plastic which are partly covered with leather. The surfaces are backed by an inhomogeneous assembly of materials and air cavities.	Not measurable in the impedance tube. Realistic consideration of mounting conditions, backing materials and air-gaps is not possible.	Unsatisfactory results. Precise measurement of these materials appears to be very difficult, due to curved shape, close interfering reflections and overall low damping.	Not applicable. Exact layer configuration and mounting is too inhomogeneous.	2 car doors (without window) were mounted in special rigid box and measured in the reverberation chamber. No reflection phase available.
dashboard						complete dashboard was mounted in special rigid box and measured in the reverberation chamber. No reflection phase available.
backshelf		The mostly leather covered backshelf is made of a thin layer of pressed wood which is mounted to a perforated metal sheet. Low frequency resonant behaviour expected due to interaction with trunk volume.	Not measureable in impedance tube. Very inhomogeneous structure with loudspeakers and head rests. Moreover, interaction with trunk volume cannot be replicated	Not applicable in-situ. Back window is to close.	Not applicable due to inhomogeneous material structure	complete backshelf was mounted in special rigid box and measured in the reverberation chamber. No reflection phase available.

Figure 9.4.: Categorization of car materials and summary of applied measurement methods (Part 2). A more detailed discussion is given in the respective subsections on the individual materials.

9.2.1. Car seats

State-of-the-art high class car seats are made of complex and inhomogeneously layered material configurations, consisting of a leather or fabric cover, a metal frame, multiple porous layers, a plastic back lining and in some cases even heating, ventilation or massage units. Figure 9.5 shows an exemplary photograph of such a seat, which was cut open to unveil the different material layers. Although the material configuration of the seat is very inhomogeneous with regard to the layering and the respective thicknesses, for the simulation model the front side of the seat was for the sake of simplicity divided into only two different areas, i.e. the seat cushion and the back rest, which were then used to assign the respective boundary conditions in the simulations.

Figure 9.5.: Photograph of cut open car seat with different material layers.

In order to cut samples from these different areas we were supplied with a single car seat and samples of different cover materials by our partner in the automotive industry. Examples of the cut-to-size material samples for the impedance tube measurements are shown in figure 9.3. Please note, that the exemplary seat shown in figure 9.5 does not exactly correspond to the seat, which was supplied for the obtainment of the seat material samples. Additionally, the samples of the single material layers were used to determine the necessary material parameters for the parameterization of the layers in the two-port network absorber model. These parameters include the density and thickness of each layer, as well as the flow resistivity of the porous layers and the hole diameter and perforation ratio of the perforated leather cover. The flow resistivities were measured according to the guidelines given in "ISO 9053:1991; Acoustics – Materials for acoustical applications – Determination of airflow resistance using the flow resistivity measurement tube" at ITA of RWTH Aachen University. Finally, the reflection characteristics of the seats were also investigated by in-situ measurements inside a car using the Microflown setup. However, although the seat, which was used to cut the material samples, was taken from the same car model as the seat, which was actually measured with the Microflown in-situ setup inside a car, it was unfortunately not possible to check if the interior of the seats was exactly the same. Possible differences might be due to variations in the equipment version of the car seats. It should also be mentioned, that only seats with a perforated leather cover were measured in the in-situ measurements.

The results of the measurements for a seat with a perforated leather cover obtained in the impedance tube, with the Microflown in-situ setup and with the two-port network model are summarized in figure 9.6. Details on the measurement results, such as number of investigated samples, measurement positions and applied averaging methods are given in the figure annotations. It can be seen that on average all applied methods predict a similar absorption behaviour of the car back rest and seat cushion. However, the comparison of the impedances reveals a considerable uncertainty of the measurement results obtained in the impedance tube and with the Microflown setup, especially in the real part of the impedance. Possible reasons for this uncertainties or even systematical errors in the measurements are given in figure 9.3.

Next, figure 9.7 shows the differences in the absorption characteristics of the seats for three different cover materials as obtained from the impedance tube measurements and the two-port network model. Despite the considerable differences between the results obtained from the impedance tube and the two-port network model which are observed for the non-perforated leather cover (which shows a strong resonant behaviour in the two-port network model, due to the airtight leather cover), the results clearly show the strong influence of the cover material on the overall reflection characteristics of the seat. This influence is presumably much stronger than the influence of material inhomogeneities within the seat, which can be guessed from the rather small variations between the Microflown measurements on different positions of the back rest and seat cushion of the perforated leather seat. It can thus be concluded, that an assessment of the average reflection characteristics of a car seat should account for the used seat cover material, while the inner material structure may be assessed by a representative sample of the back rest and seat cushion material layers.

Nonetheless, bearing in mind that the actual material layering might vary considerably among different car models, different equipment versions of the same car model or even within the backrest or seat cushion of a specific single car seat, it is clear that our assessment of the seat reflection characteristics can surely not be representative for all different kinds of seats. Problematic in this context is further, that we do not have sufficient data from different seats to actually assess the variability of the impedance and absorption data between different car seats. For the room acoustic simulations presented in this study we finally used absorption and impedance data obtained from the two-port network model, which was calculated for field incidence conditions.

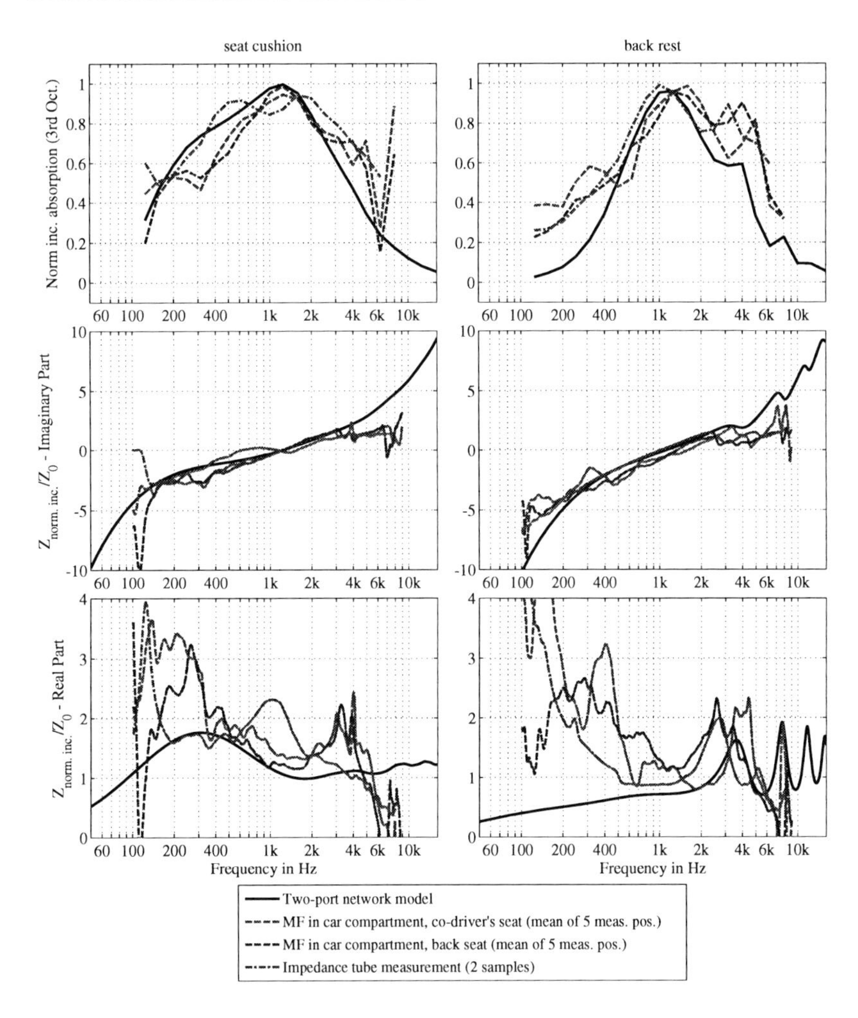

Figure 9.6.: Comparison of normal incidence impedances and absorption coefficients of the car seats with perforated leather cover.

The results are calculated from the two-port network model and measured in the impedance tube and with the Microflown in-situ method. The absorber parameters for the layers of the two-port network model were carefully measured on samples of the isolated material layers which were cut from a fully equipped car seat. The samples for the impedance tube were cut from the same car seat and the shown results were obtained by averaging the measurement results of two representative samples of the seat cushion and back rest respectively. Finally, the Microflown measurements were conducted inside a car passenger compartment on the front and back seat on the co-driver's side. The shown results for the Microflown setup were then averaged from five measurement positions which were chosen homogeneously distributed across the seat cushion and back rest respectively.

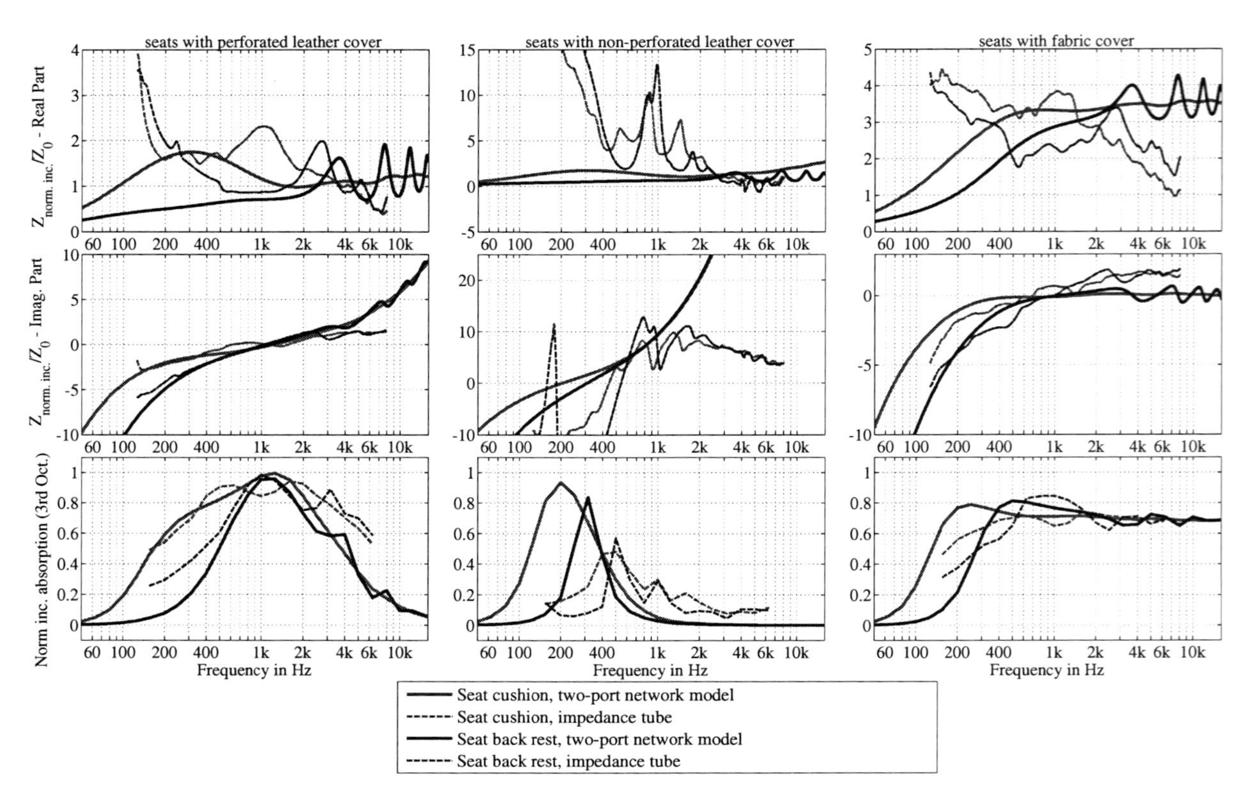

Figure 9.7.: Comparison of normal incidence impedances and absorption coefficients of the car seats for different cover materials (perforated leather, non-perforated leather and fabric).

9.2.2. Car mats and floor carpet in the foot space

The leg room in the front and back of the car is lined with a porous floor carpet with a thin airtight backing which is glued to a heavy foam of inhomogeneous thickness, which is formed to fit into the underlying sheet metal structure. In the floor areas this carpet is mostly covered with additional car mats, which are similarly made of a porous fabric with a stiff porous backing. The impedance tube samples of the floor carpet (with and without foam backing) and the car mats were cut from larger samples supplied by our industry partner and clearly show the layer structure of the material configuration (cf. figure 9.8). However, with regard to the parameterization of the two-port network model, a direct measurement of the material parameters of the individual layers in the carpet configuration was unfortunately not possible, since the layers were partly glued together and could not easily be disassembled. Therefore, only the general layer configuration and the thicknesses of the individual layers were determined directly from the material samples. The other layer parameters were then estimated within reasonable limits to give a best-possible fit with the results from the impedance tube measurements, which were carried out for the floor carpet samples alone (with and without foam backing) and for the carpet with the car mat samples on top of it. All measurements in the impedance tube were carried out with a rigid tube termination.

In addition to the impedance tube measurements and the modeling based on the two-port network approach we also tried to measure the impedance characteristics of the car mats backed by the floor carpet and thick foam layer directly in the foot space of the car using the Microflown in-situ setup. However, the first measurement attempts with the Microflown setup (not shown in this thesis) showed considerable variations around the expected absorption curve except for the highest frequencies above 8 kHz. These observed fluctuations are most likely caused by the influence of very close disturbing reflections in the foot space. Similar problems were found by Knutzen [2008] and van Gemmeren [2011], who also reported strong disruptive effects caused by close reflections when using the Microflown setup. Since due to the close vicinity of the reflecting surfaces, it did not appear reasonable to apply a time windowing approach to cancel these reflections in the postprocessing of the measured pressure and velocity impulse responses, we used additional absorptive pu foam mats to cover the surrounding reflecting surfaces (cf. figure 9.9). This made it possible to obtain reasonable results down to the lower mid-frequency range (500 Hz).

Figure 9.8 reports the results of the impedance tube and the Microflown measurements as well as the calculations based on the two-port network model. It can be seen from the left plots, which show the reflection characteristics of the rigidly backed floor carpet, that for the chosen parameterization the rigidly backed single layer Komatsu absorber model yields a reasonable fit with impedance tube measurements. This parameterization of the floor carpet was then also used for the joined layering of car mat and floor carpet, so that in this case only the parameters of the car mat layers needed to be fitted. Secondly, and more importantly it can be seen from the two-port network calculations on the right-hand side of figure 9.8 that the consideration of an exemplary heavy foam backing of 50 mm behind the thin airtight backing of the floor carpet considerably affects the low- and even mid-frequency reflection characteristics of the material configuration. Since, however, this foam backing has a highly inhomogeneous thickness it is very difficult to account for its influence on an average basis. Unfortunately, the in-situ Microflown results which are obtained for the actual built-in situation do also not give a reliable indication on the average low frequency reflection characteristics.

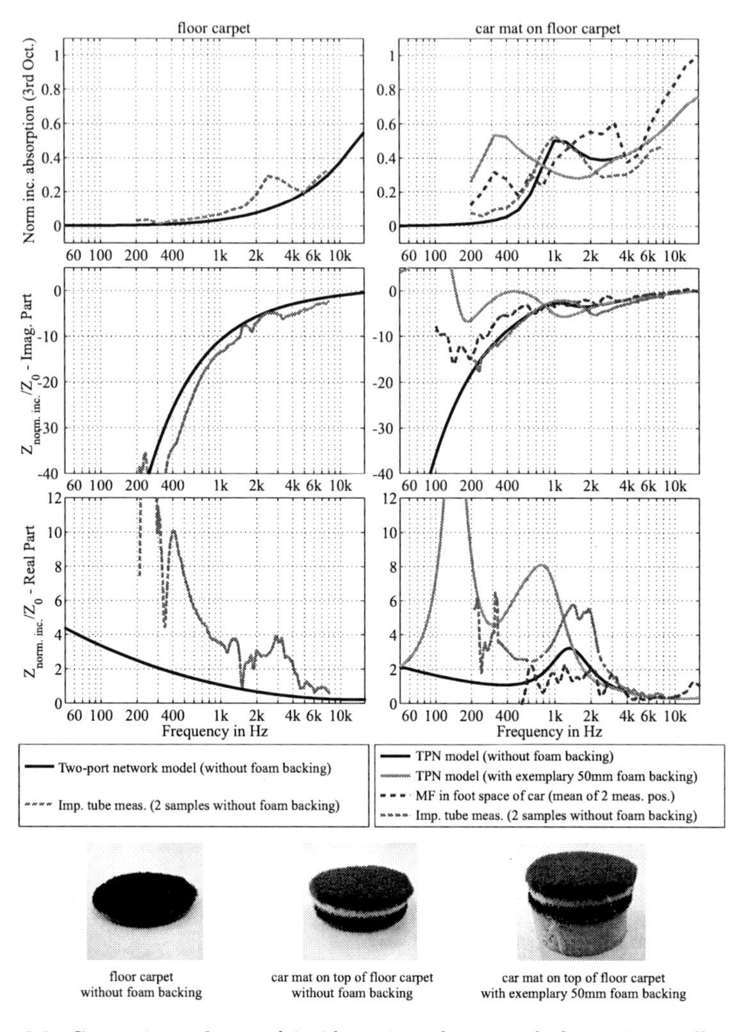

Figure 9.8.: Comparison of normal incidence impedances and absorption coefficients of floor carpet and car mats.
The results were measured in the impedance tube and with the Microflown in-situ method. The parameterization for the two-port network model was chosen to give a best-possible fit with the measured results.

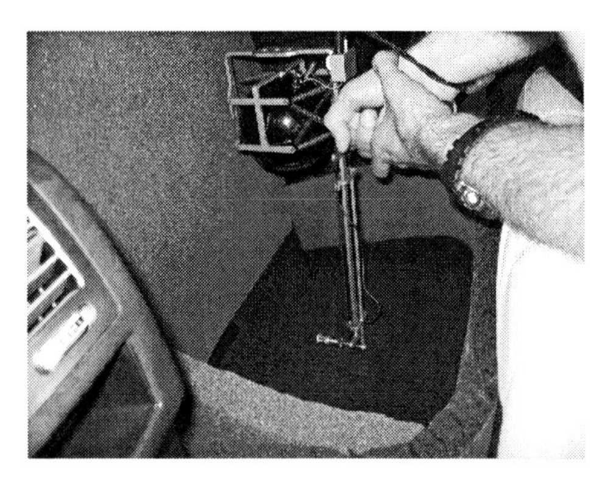

Figure 9.9.: In-situ Microflown measurement of floor carpet with car mat on top. Additional absorber mats were used to attenuate disturbing reflections from close-by objects.

For the room acoustic simulations we have therefore used the absorption and impedance data obtained from the two-port network model, where we assumed a 50 mm foam backing behind the floor carpet in the car mat areas and a rigid backing of the floor carpet for the side-areas of the legroom, since the heavy foam backing is mostly found underneath the car mats and not so much on the sides of the legroom, where the floor carpet is mostly backed by a plastic lining. Taking into account that the woofer loudspeakers are often mounted in the lower part of the door linings and therefore radiate the sound directly into the legroom, the considerable uncertainty in the determination of the low and lower mid-frequency reflection characteristics of these areas is surely a critical issue in the simulation. However, more reliable data can only be determined by a thorough measurement study based on a large set of material samples with a detailed description of the inhomogeneity of the material layers behind the thin airtight backing of the floor carpet.

9.2.3. Headliner

The headliner material in the investigated car consists of a thin porous fabric layer and a light airtight backing (see figure 9.4). In its actual built-in situation the headliner is mounted at a distance of approximately 5 cm to the metal roof sheeting of the car. As our simulation model neglects the fact that the actual headliner features some lighting elements, handle bars and in the considered case also a sliding roof, our goal is to determine representative, average reflection characteristics for the whole area of the headliner.

In a first attempt we have therefore cut out samples from a dismounted headliner and mounted these at a distance of 48 mm to a rigid termination in the ITA impedance tube. Unfortunately, as already observed in the measurements of the x-wave studio absorber, the impedance tube measurements failed to generate reproducible results regarding the resonance behaviour of the absorber configuration, due to unavoidable clamping effects of

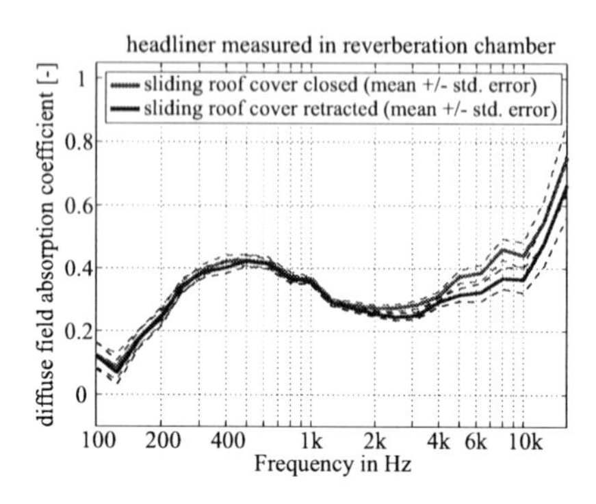

Figure 9.10.: Measurement of diffuse field absorption coefficient of car headliner in ITA reverberation chamber.
The upper photograph on the right shows how the complete headliner with metal sheeting was fitted into the rigid box. The lower photograph shows the rigid box, which was used in the reference measurements. The headliner was carefully sealed at all edges to avoid leakage effects.

the headliner sample. In a second attempt we therefore tried to determine the reflection characteristics directly in the car using the Microflown in-situ setup. However, similar to the in-situ measurements of the car mats, these measurement results also showed strong disruptive effects. While these can again be partly attributed to close interfering reflections from windows and head rests, a comparably strong error contribution might be due to the unevenness and/or the slightly curved shape of the headliner. In this context van Gemmeren [2011] has shown that even in the absence of any surrounding reflections, good results with the Microflown setup can only be expected if the considered material samples are sufficiently flat in a reasonably large area around the measurement position. Due to the unsatisfactory results obtained in the impedance tube and with the Microflown setup, the impedance and absorption characteristics were also estimated with the two-port network model. Similar to the car mats and floor carpet the parameterization of the material layers was again chosen within a reasonable parameter range and with an eye on the measured data. The results are shown in figure 9.11.

In a final step we have also attempted to measure the average absorption characteristics of the headliner in the reverberation chamber. The whole headliner with metal roof sheeting was therefore carefully fitted into a rigid box, which was damped on the inside. The average absorption coefficient was then determined using the method described in section 5.4. In order to account for edge effects we conducted the reference measurements with the same rigid box in the room, only that this time it was sealed with a flat rigid panel (cf. figure 9.10). The results of the reverberation chamber absorption measurements are also given in figure 9.10. These results were then finally used in the room acoustic GA simulations. For the FE simulations a real-valued impedance was calculated from the measured field incidence absorption values using equation 5.2.

It is important to mention at this point that the negligence of the phase information in the impedance of the headliner, presumably causes noticeable errors in the FE predictions of the low frequency transfer function inside the car. However, due to the considerable uncertainty in the parameterization of the two-port network model as well as in the Microflown and impedance measurements, we have decided to use the reverberation room data in order to at least get a realistic estimation of the average damping introduced by the headliner surface in the car compartment.

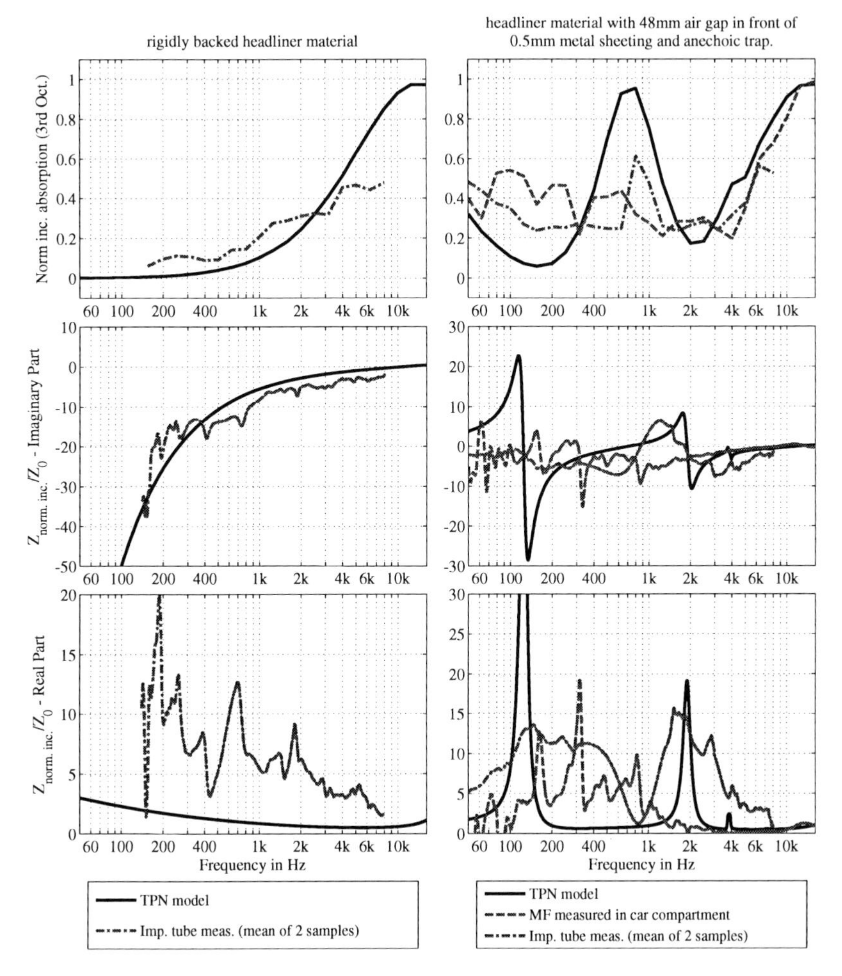

Figure 9.11.: Comparison of normal incidence impedances and absorption coefficients of headliner.

The results were measured in the impedance tube and with the Microflown in-situ method. The parameterization for the two-port network model was chosen to give a best-possible fit with the measured results.

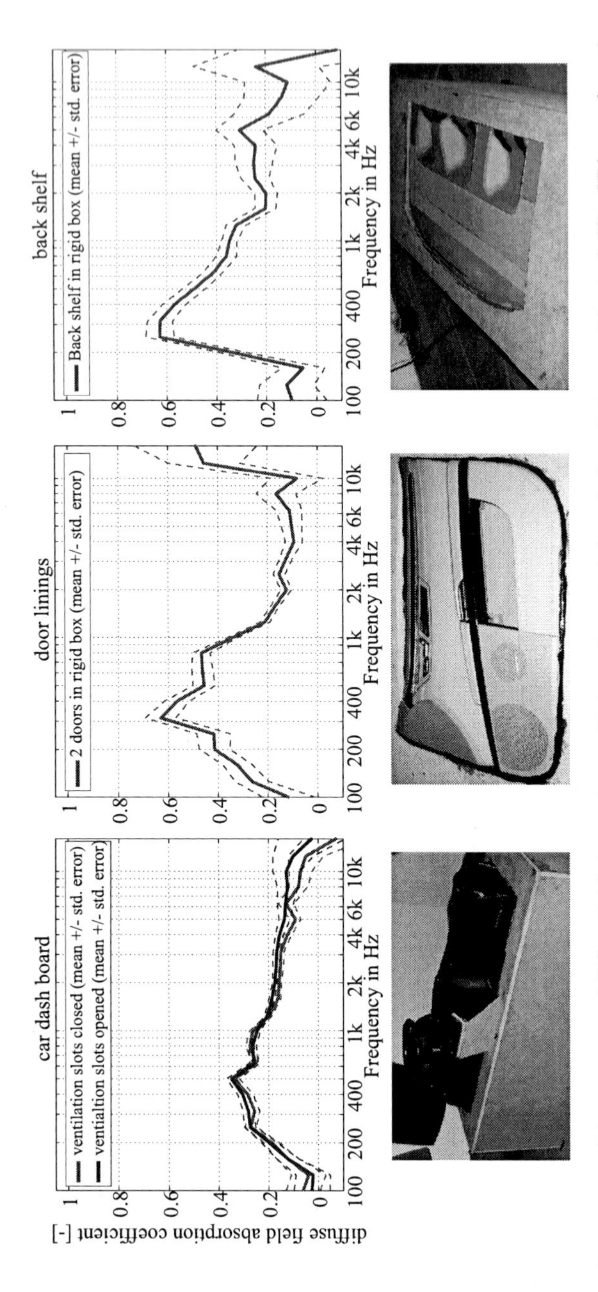

Figure 9.12.: Measurement of diffuse field absorption coefficient of car dash board, door linings and back shelf in ITA reverberation chamber.
The photographs in the lower part of the figure show how the components were fitted into special rigid measurement boxes. For the reference measurements the top plates with the car components were removed and replaced with a thick flat MDF plate. All car components were carefully sealed at all edges to avoid leakage effects.

9.2.4. Dashboard, door lining and back shelf

Due to the immense inhomogeneity regarding their shape and material configuration the car dashboard, back shelf and door lining were not accessible to an investigation in the impedance tube or with the two-port network model, since representative measurement samples could not be determined for these structures. Moreover, an alternative attempt to measure the acoustic reflection characteristics in-situ using the Microflown setup also failed due to the aforementioned problems related to close interfering reflections and insufficiently large flat areas of the materials under investigation. While these two factors generally suffice to completely ruin any in-situ impedance measurement, we want to add for the sake of completeness, that the comparatively low absorption of the considered materials further complicates the in-situ measurements. As described by van Gemmeren [2011] the problems when measuring low absorptive samples can be explained by an increased sensitivity of the method to slight errors in the experimental setup as well as to any sort of noise.

In order to still obtain measured average absorption characteristics for the dashboard and door lining we have again resorted to the reverberation room method described in section 5.4. The measurements were carried out in the exact same way as for the headliner. Figure 9.12 shows the measurement results and photographs of the material samples which were carefully fitted into the measurement boxes and sealed at the edges. The so obtained absorption characteristics and the deduced real-valued impedances were finally used in the room acoustic simulations of the car compartment. Again, the negligence of the reflection phase might have a detrimental influence on the FE simulation results. However, neither a direct measurement of the complex impedance values nor a prediction with the two-port network model appeared feasible for the considered materials.

9.2.5. Windows

Although the windows in the car can be considered as almost acoustically rigid except for the lowest frequency bands, the two-port network model was used to realistically capture the small absorption contribution of these surfaces at low frequencies. In the FE domain the windows were therefore modeled by an effective mass using equation 8.1 with a free field termination ($Z_{\text{term}} = Z_0$). In the GA domain a simple surface mass model was used instead (cf. section 8.4.3 for details). Both the surface impedance and absorption characteristics were calculated for field incidence conditions. The necessary material parameters for the effective mass model were estimated from online available data tables on the dynamic behaviour of glass. Although the used parameterization might not be completely accurate, the influence of this error on the room acoustic simulation results is considered to be very low due to the overall small contribution of these surfaces to the total absorption in the car compartment.

9.2.6. Other surfaces

The remaining surfaces in the car comprise the center console, the back sides of the seats and the areas underneath the front seats, whose reflection characteristics could not be determined by direct measurement or model calculation. Due to the similarity of the materials used for the center console and the dashboard, the absorption characteristics, which were measured in the reverberation chamber for the dashboard, were also assigned to the center console. Moreover, since the back sides of the seats are not expected to strongly contribute to the overall damping in the car compartment their absorption coefficient was set to an estimated broadband value of 0.15. The acoustic characteristics of the areas underneath the seats appear even more difficult to estimate. Due to the lack of any measured data the absorption coefficient was set to a somewhat arbitrary broadband value of 0.3. In both cases an according real-valued impedance was calculated on the basis of the estimated broadband absorption value.

9.3. Source conditions

The two car models that were investigated in the course of this study are equipped with sound systems consisting of 14 (model 1) and 16 (model 2) loudspeakers with a number of 5 and 3 different driver units, respectively. The required simulation input data was obtained from velocity, pressure and directivity measurements where the single driver units were mounted flush into the hard reflective floor of the hemianechoic chamber as described in section 6.1.1, figure 6.1 (a). Exemplary photographs of the measurements of the car loudspeakers are shown in figure 9.13.

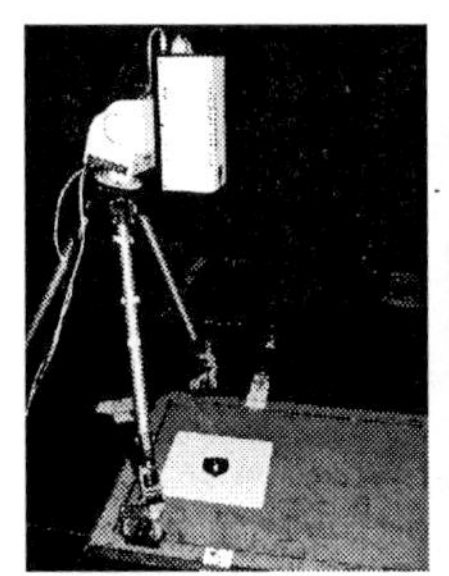

laser vibrometer measurement of membrane velocity @1V for FE source representation

Free-field measurement of on-axis pressure transfer function @1V, 1m for GA source representation

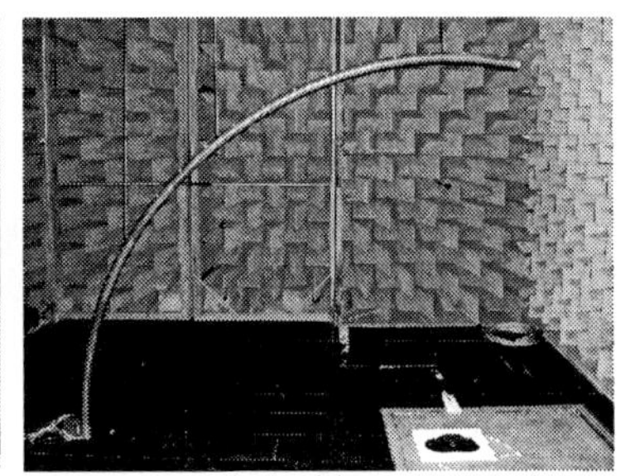

Anechoic directivity measurement of rotation symmetric loudspeaker on quarter circle for GA source representation

Figure 9.13.: Measurement setups used for the determination of the source characteristics of the car loudspeakers.
All measurements were conducted on the individual, dismounted driver units which were mounted flush into the hard reflective floor of the hemianechoic chamber.

In addition to the measurements idealized membrane velocities of the loudspeakers were calculated using Thiele-Small data that was supplied by our industry partner and which was obtained using the Klippel R&D System[4]. The so obtained membrane velocities $\underline{v}_{\mathrm{m}}(f)$ at 1 V input voltage at the voice coils were then used to predict the free field pressure transfer function $\underline{p}_{\mathrm{ff}}(f)$ at 1 V, 1 m under the assumption that the loudspeakers work as ideal piston sources in an infinite baffle. Moreover, the directivity $D(\theta)$ of the loudspeakers was estimated on the basis of the analytic formula for a piston in an infinite baffle:

$$\underline{p}_{\mathrm{ff}}(f) = \rho_0 c\, \underline{v}_{\mathrm{m}}(f) \left(e^{-jkr} - e^{-jk\sqrt{r^2+a^2}}\right), \quad \text{with} \quad r = 1\,\mathrm{m}, \tag{9.1}$$

$$D(\theta) = 2\frac{J_1(ka \sin\theta)}{ka \sin\theta}, \tag{9.2}$$

where a is the effective radius of the membrane and θ is the angle of radiation ($\theta = 0$ is the main direction of radiation).

For our room acoustic simulations presented in section 9.4 we exclusively use the measured diaphragm velocity, free field pressure and directivity data obtained in the anechoic chamber. For the sake of brevity, we will however not report the results of all the loudspeaker measurements conducted within the study. Instead, we want to give a more general discussion on the applicability of the presented methods which were used to determine the source data for the simulations in the car compartment. The plots presented in this paragraph thus show exemplary results of car loudspeakers that were all investigated in the course of the project. The plots and the corresponding loudspeakers were chosen for illustrative reasons to corroborate the discussed problems and aspects. Please note that not all plots refer to the loudspeakers that were used in the selected simulations in section 9.4. In particular, two questions shall be discussed; (a) does the measurement of the source data with the measurement setups shown in figure 9.13 capture all relevant characteristics of the loudspeakers with regard to their real built-in situation in the car and (b) how big are the errors of the analytically predicted data based on the formulas given in equations 9.1 and 9.2.

With regard to the first question it can be stated that the used measurement setup obviously neglects any diffraction effects that are caused by the actual built-in situation of the loudspeaker. Such diffraction effects are for example caused by the fact that the loudspeaker drivers are generally not mounted flush into the door liners, the dash board or the back shelf. Instead the loudspeaker drivers are generally mounted in a small recess with a grid cover and may even be slightly tilted within this recess (see figure 9.14). While this surely has noticeable effects on the mid- and high frequency directivity function of the loudspeaker, the free field pressure transfer function might also be affected to a certain extend. For example in the case of the tweeter in the mirror side mountings we have found a strong comb-filter at about 10 kHz due to a small plastic tube between the loudspeaker diaphragm and the grid cover of the loudspeaker. For the determination of the GA source data it would therefore be preferable to measure the pressure response and directivity data for the fully assembled loudspeaker in the recess, with grid cover and if present also with the corresponding back side cavity. However, since we were only supplied with the single driver units, this was unfortunately not possible in the present study.

[4] http://www.klippel.de/our-products/rd-system.html

Figure 9.14.: Low- and midrange driver units in recess in lower part of front door (with and without grid cover).
The photographs clearly show that the loudspeakers are not mounted flush into the door liners as in the anechoic measurements shown in figure 9.13. Instead, the loudspeakers are mounted in a recess with protective grids, which will influence the loudspeaker directivity and possibly also the loudspeaker transfer function at higher frequencies.

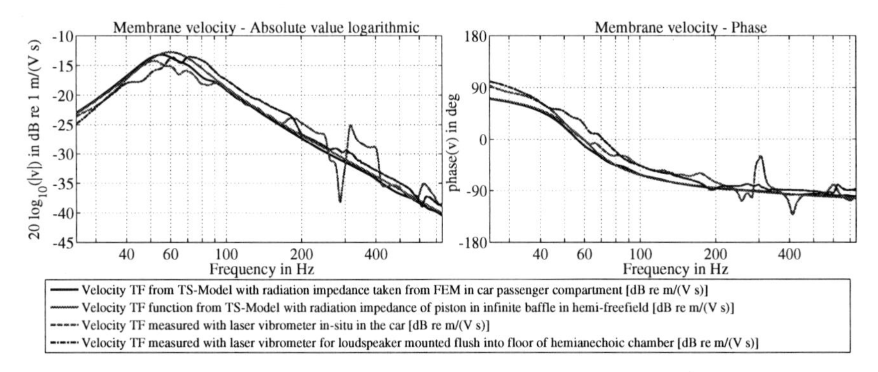

Figure 9.15.: Membrane velocity of car woofer loudspeaker measured and calculated from Thiele-Small model.

On the other hand the determination of the diaphragm velocity function is mostly independent of the exact mounting configuration, the recess and the grid cover of the loudspeaker. However, although we have found in section 7.3 that the room sound field generally has no considerable influence on the diaphragm velocity, figure 9.15 reveals considerable differences between velocity transfer functions measured directly inside the car and those measured in the free field (with the measurement setup shown in figure 9.13). Additionally figure 9.15, plots the velocity curves obtained with the Thiele-Small model using (a) the free field radiation impedance and (b) the radiation impedance obtained from FEM simulations in the considered car passenger compartment. While the velocity TF measured in the free field agrees well with the predictions of the Thiele-Small model (irrespective of the used radiation impedance in the equivalent network model), the velocity TF measured in-situ in the car shows significant deviations in the frequency range

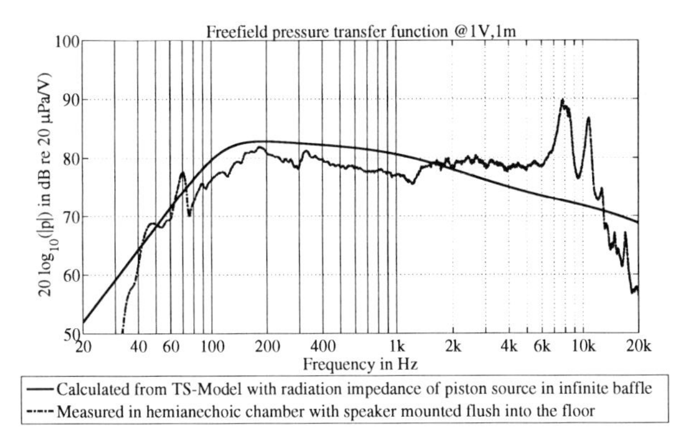

Figure 9.16.: Free-field pressure transfer function of car midrange loudspeaker measured and calculated from Thiele-Small model.

from 200 to 400 Hz, compared to both the Thiele-Small predictions and the laboratory measurements under anechoic conditions. These differences are mostly likely caused by strong modal effects in the closed loudspeaker cavity on the backside of the loudspeaker. This notion is also supported by our results in section 7.3.3, where we show that the diaphragm velocity can in fact be influenced by strong undamped modes in the rear loudspeaker cavity. Thus, although the diaphragm velocity is not influenced by the sound field in front of the loudspeaker and thus need not be measured in its actual built-in situation in the car, it is important to consider the real rear cavity of the loudspeaker box in the velocity measurements. As mentioned above this was unfortunately not possible in the course of this project.

It should finally be mentioned that the analytically calculated pressure and directivity functions based on the piston-in-baffle model (see equations 9.1 and 9.2) obviously struggle with the same shortcomings regarding the non-consideration of the actual built-in situation as the anechoic measurements of the single loudspeaker units mounted flush into the floor. Additional problems of the idealized piston-in-baffle model for use as GA simulation input data have already been discussed in section 6.1.2 and are mostly due to modal effects on the loudspeaker diaphragm, which cannot be considered by an electro acoustic network model based on lumped components. Figure 9.16 gives a quantitative assessment of this error based on the comparison of free field pressure transfer functions which are (a) calculated from equation 9.1 (where the diaphragm velocity is calculated from the equivalent network diagram of the loudspeaker) and (b) measured using the measurement setup shown in figure 9.13, where the loudspeaker is mounted flush into the floor of the hemianechoic chamber. As expected we observe considerable high frequency deviations (above 5 kHz) due to pronounced diaphragm eigenmodes. On the other hand the differences below 100 Hz are caused by low frequency eigenmodes in the anechoic chamber and are not due to spatial variations of the velocity on the loudspeaker membrane.

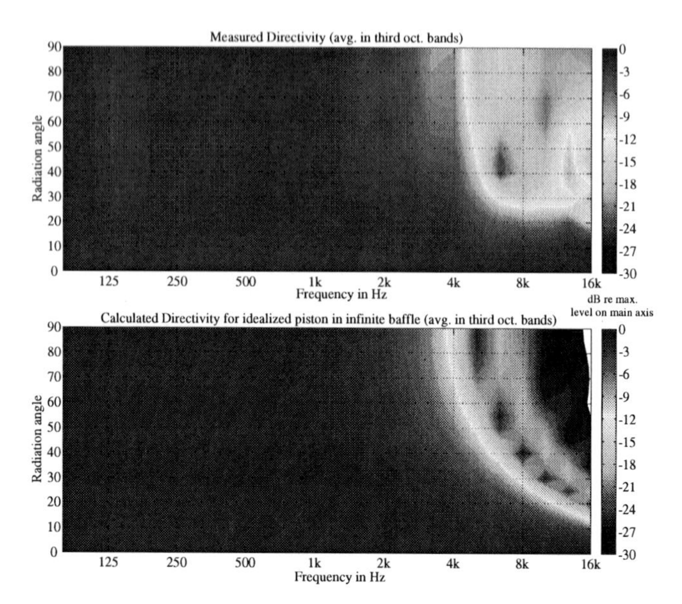

Figure 9.17.: Comparison of measured and calculated directivity functions for car midrange driver.
The measurements were carried out in the ITA anechoic chamber (cf. figure 9.13) and the calculations are based on formula for a piston in an infinite baffle (cf. equation 9.2).

For the above mentioned reasons it is therefore generally not recommended to use an idealized pressure TF calculated from equation 9.1, when high requirements on the GA simulation accuracy exist. While the above made reasoning similarly applies to the directivity data, the application of equation 9.2 might be admissible from a practitioners point of view since (a) no laborious directivity measurements are required, (b) the directivity is not so sensitive to modal effects on the diaphragm (at least if the membrane is rotation symmetric) and (c) despite the negligence of the actual built-in situation the general constriction of the sound radiation to the normal direction should in a first approximation be captured sufficiently well by this simple prediction formula. Figure 9.17 shows a comparison of the directivity maps of a rotation symmetric loudspeaker measured in the floor of the ITA anechoic chamber and calculated from the idealized piston-in-baffle formula. An investigation of the influence of the actual mounting conditions of the loudspeaker on the directivity function could unfortunately not be conducted in the course of this study.

Summing up, we have identified non-negligible issues regarding the applied methods that were used to obtain the source data for the combined FE-GA simulations and possible ways to overcome these problems have been suggested. Although it was unfortunately not possible to obtain better source data within the course of the project (since only the single loudspeaker drivers were available for the measurements), we give the following recommendations for the determination of the simulation input data of loudspeakers used in a car passenger compartment:

- **Diaphragm velocity for FE simulations:** For the determination of the excitation velocity a direct measurement using a laser vibrometer under in-situ or laboratory conditions is recommended. For best possible results a spatial scan of the loudspeaker diaphragm should be conducted and the so obtained spatially varying velocity (or alternatively the surface-averaged velocity) should be assigned to the membrane surface in the FE simulation. The consideration of the spatial variation of the membrane velocity is especially important for frequencies above the piston range of the considered loudspeaker (cf. section 6.1.2). The velocity measurement does not have to be carried out in-situ in the car. However, if it is carried out on a dismounted loudspeaker unit, it is important to consider the rear cavity of the loudspeaker in the measurement in order to account for the possible influence of cavity modes (if there exists a closed rear cavity). If the loudspeaker works on a rather undefined rear cavity, which is e.g. the whole cavity behind the door liner or the whole trunk in the case of the subwoofer on the back shelf, it might be necessary to consider the whole door/trunk for an exact measurement of the excitation velocity.
- **Free-field pressure transfer function for GA simulations:** The pressure transfer functions and the directivity functions need to be measured under free field conditions in an anechoic or hemianechoic environment. In order to account for diffraction effects caused by the mounting conditions of the loudspeaker (recess, grid cover) the measurement should be carried out for the loudspeaker in its actual housing. If two loudspeakers in the car are mounted in very close vicinity to each other, e.g. a tweeter and a low- to mid-range driver in a door, these units can also be considered and measured as a single sound source. This is however only reasonable if both loudspeakers are fed with the same input signal and the crossover between the two loudspeakers is fixed and considered in the measurement. The directivity measurement should be carried out on a sufficiently fine grid discretization for the hemisphere above the loudspeaker. Finally, special care should be taken in the room acoustic simulation setup when placing and orienting the sources in the car model (with regard to the view and up vector), in order to account for possible tilting of the loudspeakers in the real mounting conditions.

While following these recommendations will give high quality input data for the FE domain and also improved data for the GA domain, some general concerns with regard to the source description in the GA domain remain. These concerns are attributed to the close proximity of reflecting surfaces to the loudspeakers which might cause near field effects that influence the sound radiation pattern. This is especially problematic for the tweeter loudspeakers in the side mirror mountings (between the wind shield and side window) or for the midrange loudspeakers on the dash board and back shelf which radiate the sound directly into the very close adjacent windows. A quantitative assessment of the influence of the error caused by these close reflecting surfaces can however not be given on the basis of the data obtained during the project.

9.4. Comparison of simulation and measurement results

The current section presents the comparison of simulation and measurement results obtained for the two car models. As already mentioned earlier, the main goal of the project was to generate simulation-based, binaural auralizations of the sound field in a car compartment excited by the loudspeakers of the car sound system. However, since the simulation setup used for these simulations includes many uncertainties with regard to the boundary, source and receiver representation, which can hardly be separated in the final simulation results, we have run additional simulations which allow us to better quantify how the different influencing factors contribute to the observed differences between measured and simulated room transfer functions.

In particular, we have tried to separate the impact of errors in the source and the boundary representation, by conducting monaural measurements and simulations using the small dodecahedron loudspeaker, which was already used in the model room study (see section 7.2), as the sound source in the car compartment. Since this dodecahedron loudspeaker can be described in good approximation as an omnidirectional point source in the FE and GA domain (cf. section 7.2) the observed differences between measurement and simulation results can, at least in the FE domain, be almost exclusively attributed to errors in the geometry and the boundary conditions. In the case of the GA domain additional errors are introduced by the negligence of diffraction and near field effects as well as general limitations of the chosen energy-based geometrical acoustics approach (see section 4.2). Nonetheless, the simulations with the dodecahedron source take at least one unknown out of the equation.

Moreover, it has been investigated if significant differences in the quality of the monaural and binaural simulation results can be observed. Possible reasons for such differences have to discussed separately for the FE and GA domain. While in the FE domain we are mostly concerned with the question, if a dummy head and torso model needs to be included into the FE mesh (which is in the end only a question of the required time and effort needed for the simulation preparation), much more difficult questions arise in the GA domain. These questions are related to the absence of an actual receiver geometry (head/torso) and the use of far-field measured HRTF data in the binaural GA simulations. While these simplifications appear surely adequate in larger rooms, their impact on the simulations in the very small car compartment is difficult to assess. The results of the comparative simulation and measurement studies are given in the following subsections.

9.4.1. Monaural results with small dodecahedron source (car model 1)

For the present investigation room transfer functions were measured inside the empty compartment of car model 1 from the small dodecahedron source which was placed on a stand on the back seat as shown in figure 9.18 (b) to 4 receiver positions which were located at head height in front of the head rests of all four main seat positions. Figure 9.18 (a) shows how the used Sennheiser KE4 microphone capsules were hung from the head liner. In the simulation setup we tried to replicate the source and receiver position to the best

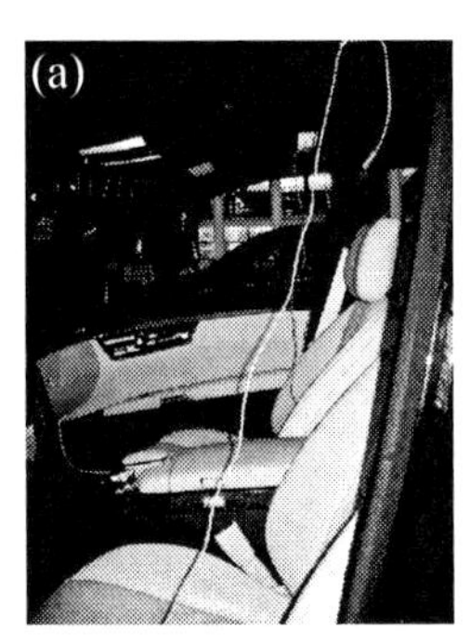

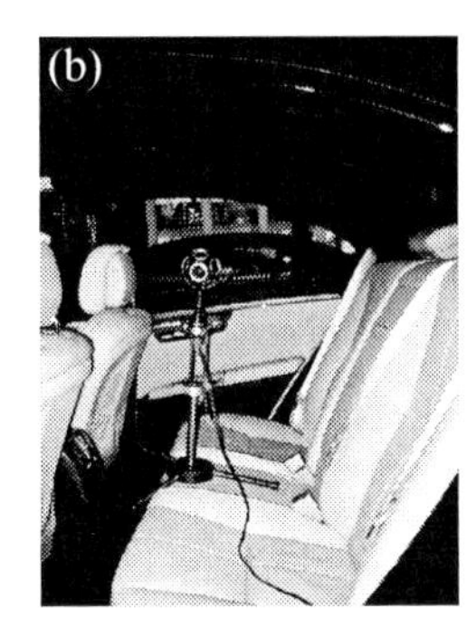

Figure 9.18.: Configuration of small dodecahedron loudspeaker source on back seat and measurement microphones hung from headliner in car model 1.

possible extend. The source was modeled as an omnidirectional point source in the FE and GA domain. For direct comparability of the simulation and measurement results, we used the same normalization of the source strength as described in section 7.2.5, which means that the measured and simulated transfer functions were normalized to the free field pressure response at 1 V, 1 m of the respective sound source (i.e. the small dodecahedron in the measurement domain and the ideal point source in the simulation domain). All boundary conditions were chosen as described in section 9.2. The FE simulation was run using the refined 40 mm mesh, which allowed to calculate the FE results up to 3 kHz. A comparison of the simulated and measured room transfer functions, averaged band energies and reverberation times for the microphone on the driver's position is given in figure 9.19. Additionally a comparison of spectrogram plots generated from the measured and simulated room impulse responses is given in figure 9.20.

It can be seen that despite the obvious differences in the fine structure of the measurement and simulation results, the overall frequency dependent energy distribution and reverberation characteristics are generally well captured by both the FE and GA simulation in their respective validity ranges (except for the slightly overestimated mid- and high frequency reverberation times in the GA domain). Taking into account the considerable uncertainty in the description of the boundary conditions a perfect match of measured and simulated results was surely not expected. With regard to the FE results it would therefore not make much sense to speculate about the reasons for each little difference in the modal structure of the low frequency sound field. Especially not, since it was already shown in section 7.1 that the negligence of the phase information in the acoustic reflection characteristics (as done here for the headliner, dashboard, backshelf and door linings) may significantly change the modal structure of the low frequency sound field. However, it is important to note that thanks to the well-controlled source and the high quality geometric room model, the observed differences between measurement and FE simulation can almost exclusively be attributed to errors in the impedance boundary condition. A more precise prediction of the low frequency room modes can thus only be achieved with more realistic boundary conditions, which appears difficult to achieve considering the high complexity of some of the used material structures. On the other hand, the observed

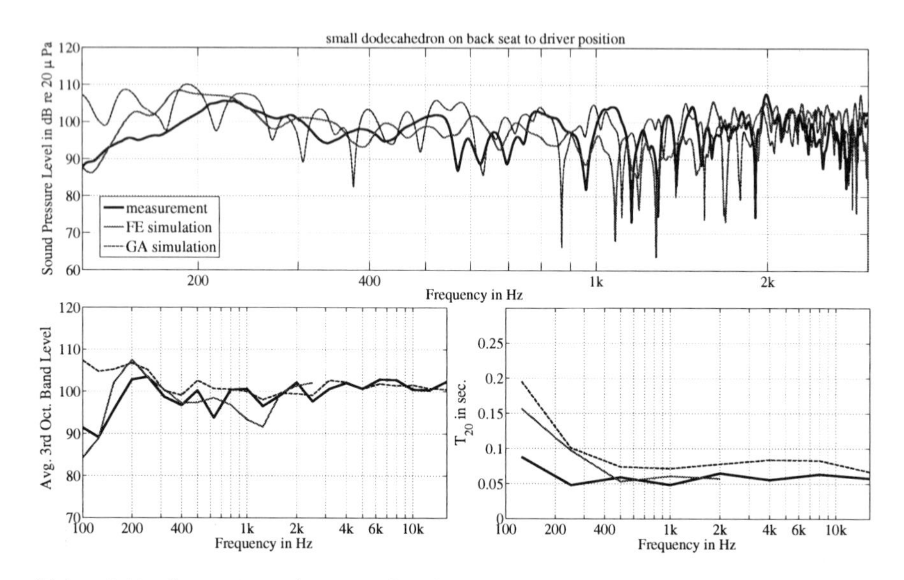

Figure 9.19.: Comparison of measured and simulated monaural room transfer functions, averaged band energies and reverberation times for car model 1 with the small dodecahedron sound source on the back seat.
The results are given for the driver's position. The microphone was positioned approximately at the driver's head position.

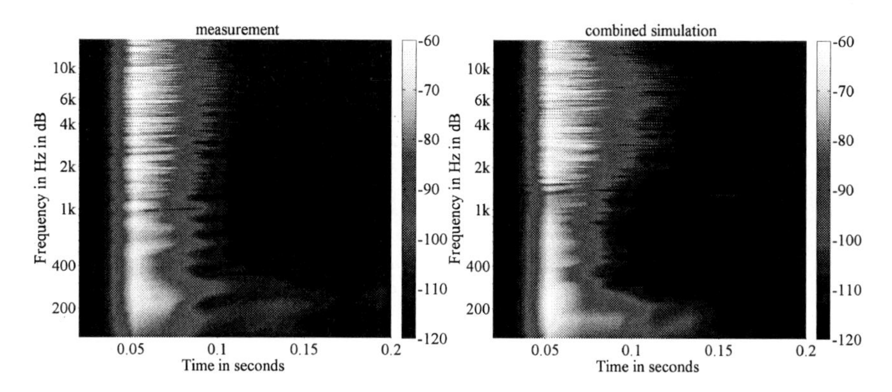

Figure 9.20.: Comparison of spectrograms obtained from measured and simulated monaural room transfer functions for car model 1 with the small dodecahedron sound source on the back seat.

differences in the high frequency reverberation characteristics between the measurement and the GA simulation, appear more unproblematic since an iterative adjustment of the absorption characteristics would easily improve the fit between measured and simulated RTs. The presented results thus represent a best possible a-priori prediction (i.e. without knowledge of the measurement results) of the sound field in the car compartment using a well-controlled sound source.

Another interesting aspect of the results is, that for undisclosed reasons the FE simulation shows a noticeable underestimation of the average band energy levels in the frequency range from 800 Hz to 1.25 kHz. In contrast to this, the geometrical acoustics simulation fits the measured results very well in this frequency range. This is somewhat surprising considering that the GA simulation neglects all diffraction effects at the driver's seat and head rest, while the FE simulation fully covers these effects. While this result does surely not indicate that the GA simulation is generally more accurate than the FE simulation in this frequency range, it raises the question if the FE simulation can actually deliver a benefit over the GA simulation in the frequency ranges with considerable modal overlap, at least in the considered case with high uncertainty regarding the boundary conditions. This means that despite the fact that the FE result captures the physics of the sound propagation very well, its results at high frequencies exhibit a similarly stochastic character as the GA results, since the boundary condition errors lead to a stochastic superposition of the overlapping modes. However, this question can surely not be answered on the basis of a single simulation result.

It is finally interesting to note the differences in the reverberation times obtained between the FE and the GA results in the mid-frequency range. This result is not expected since the boundary description in the FE and GA domain was determined from the same consistent input data (either from the two-port network model or from the reverberation chamber absorption measurements). Irrespective of the measured reverberation characteristics, the GA method should in its range of validity yield the same reverberation characteristics as the corresponding FE simulation. The observed discrepancy may thus be interpreted as an indicator that the GA simulation does not cover all relevant sound propagation aspects in the considered frequency range.

9.4.2. Monaural results with car loudspeakers (car model 1)

In a next step we have carried out analogous simulations and measurements as above, only this time with the loudspeakers of the car sound system as sound sources. Although the measurements could not be conducted in exactly the same car as used for the measurements with the dodecahedron source, the measurements were carried out in a car of the same model type and with the same equipment version. The measurement microphones (Sennheiser KE4 capsules) were positioned at approximately the same receiver points as used in the measurements with the dodecahedron source. The individual car loudspeakers were accessed using a special breakout box (supplied by our industry partner) which allowed us to directly connect to each loudspeaker without routing our measurement signals through the car PA system. The room impulse responses were then obtained from sweep measurements which were played back consecutively through each car loudspeaker and recorded with the measurement microphones. For the simulations we used exactly the same simulation setups as before (40 mm mesh) except for the now different source positions and source characteristics. The simulations were run individually for three representative sources, i.e. the woofer in the front door on the driver's side, the midrange driver in the center position on the dashboard and the integrated loudspeaker unit consisting of the low- to midrange driver and the tweeter in the back door on the passenger's side. The low- to midrange driver and the tweeter in the back door acted as a single source unit in the breakout box and could therefore not be measured separately. Instead, they were fed simultaneously with the same sweep signal in our measurements. In contrast to that, the simulations were run individually for both loudspeakers and then added in the post-processing. Unfortunately, since the measurements were not conducted with a fully calibrated output and input chain, the absolute levels of the measurements needed to be adjusted by hand to get a best possible fit with the results of the FE and GA simulations, which are inherently calibrated to each other (cf. section 6). Due to the considerable difference in the frequency characteristics of the measured and simulated results this was not unambiguously possible. By comparing the average levels of the measurement and simulation results for all single loudspeakers, we have finally deduced a single correction factor for all measurement results, which best possibly fitted the simulated levels on an average level. Since the measurement chain was exactly the same for the measurement of each loudspeaker, a single correction factor suffices for all measurements.

A comparison of the simulation and measurement results is plotted in figure 9.21. Since in the real car audio system the audio signals are bandpass filtered in a digital mixing unit before being routed to the different loudspeakers, the simulations and measurements were also band-passed using similar filters as in the mixing unit in the real car audio system. The cut-off frequencies for the bandpass filters were supplied by our partner in the automotive industry. The results are generally similar to those obtained for the small dodecahedron source. With regard to the energy distribution and reverberation characteristics most of the simulated results show an acceptable and in some cases even very good match with the measured results. Moreover, the FE simulations partly capture the modal structure of the low frequency sound field below 800 Hz quite well, although the quality factors of the room eigenmodes are not always predicted right.

By going more into detail it is striking that the GA simulation of the woofer loudspeaker in the left front door shows consistently higher band energies than the corresponding FE results. This discrepancy might be explained on the one hand by the different source representations in the FE and GA domain, since in the FE mesh the loudspeaker membrane is modeled within the actual recess shown in figure 9.14 while in the GA model the source is modeled as a point source in close distance to the doorliner. This point source uses the directivity data which was measured on a hemispheric grid, when the loudspeaker was mounted flush into the rigid floor of the hemianechoic chamber. Thus the source does not radiate any sound particles to the back side, i.e. into the doorliner. Moreover, the woofer loudspeaker itself has some peculiarities (voice coil on the front side, complex mounting in recess, unknown closed rear cavity) which make it difficult to model in both simulation domains. Finally, another possible reason for the poor agreement of the energies in both domains is given by the complicated radiation conditions of this loudspeaker since it primarily radiates into the heavily damped footspace.

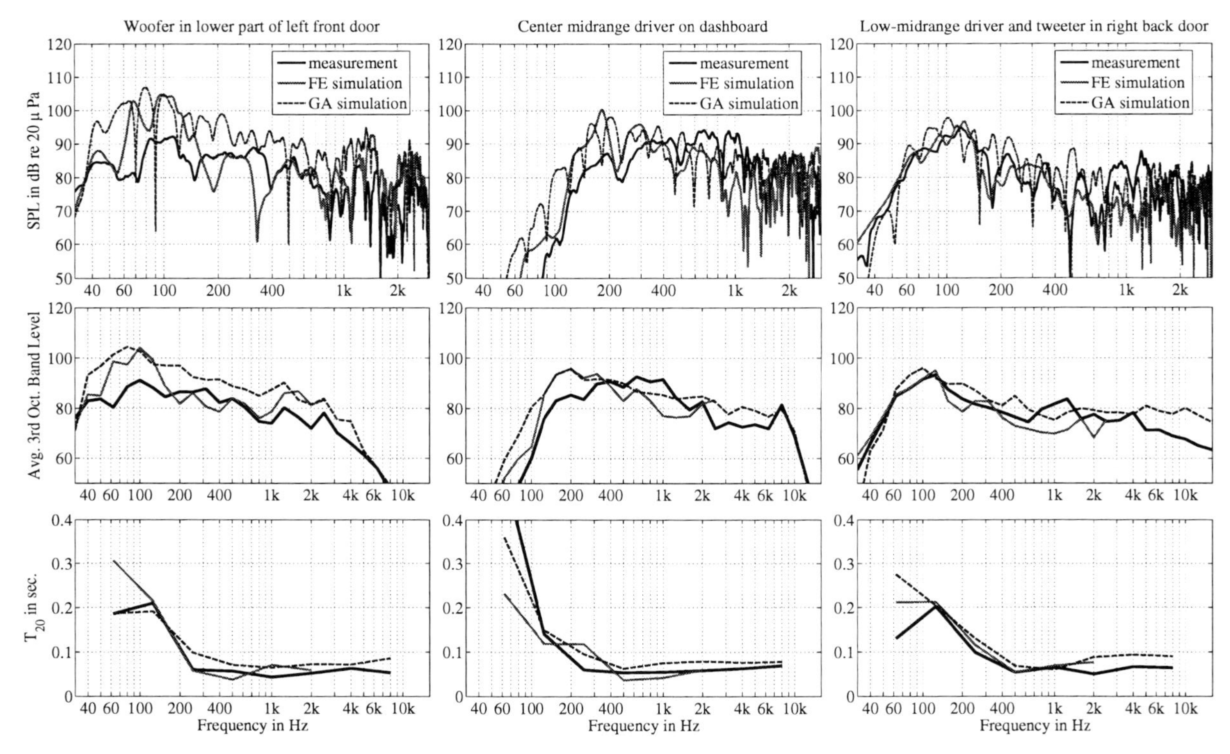

Figure 9.21.: Comparison of measured and simulated monaural room transfer functions, averaged band energies and reverberation times for car model 1 with 3 different car loudspeakers as sound sources.
The results are given for the driver's position. The microphone was positioned approximately at the driver's head position.

It should also be mentioned that the piston range of the woofer and the low- to midrange driver does only extend up to approximately 800 Hz, which makes the use of the piston models with a homogeneous surface velocity questionable for higher frequencies in the FE domain. In particular, we found that the high observed levels in the FE simulation of the woofer loudspeaker at frequencies above 1.5 kHz are caused by broad peaks in the measured membrane velocity transfer function, which was assigned to the piston surface in the FE model. Measurements at other positions on the same loudspeaker membrane did indeed reveal that the membrane velocity is no longer homogeneous in this frequency range. This problem could be alleviated by using either full spatial scans of the membrane velocity as input data for the FE domain or at least surface averaged velocity data instead of the single-point measured data which was used in the present study. Summing up, it can be concluded that the use of the actual car loudspeakers with their complex mounting conditions and special constructions introduces a further uncertainty factor into the simulations, which is difficult to quantify. It is therefore important to carefully check for each loudspeaker in how far the used simulation source model deviates from the real loudspeaker source in its specific built-in situation. Possible flaws of the source determination methods applied in this study have been discussed in detail in section 9.3.

9.4.3. Binaural results with car loudspeakers (car model 1)

The present subsection presents the binaural simulation and measurement results obtained for car model 1 with the loudspeakers of the car sound system as sound sources. The measured room transfer functions were obtained in the same measurement sessions as the monaural results presented in the previous section. Instead of the monaural KE4 microphones we used a *HEAD Acoustics* artificial head equipped with two microphones at the ear canal entries of the left and right ear. For the measurements the head was consecutively placed at a realistic height on each seat using a special stand. Except for the head no further occupants or equipment were present in the car compartment during the measurements. Except for the replacement of the measurement microphones with the dummy head the measurements were carried out in the same way as described in the previous sections using sweep excitation signals fed into the individual loudspeaker units with the special breakout box.

As already mentioned earlier the inclusion of a binaural receiver in the room acoustic simulations is not unproblematic both in the FE and GA domain. In order to investigate the influence of a head and shoulder model on the low frequency sound field, we have therefore run FE simulations using the mesh with included dummy head model (see section 9.1) on the driver's seat and compared the results at the left and right ear position with those obtained on the same seat using the model without the dummy head, where the monaural receiver is placed at the position which corresponds approximately to the centre of the dummy head in the model including the head. The simulations with the dummy head model were run using exactly the same car geometry (except for the presence of the head) and the same source and boundary conditions as in the simulations without the dummy head. Both simulations were run using the meshes with an average element size of 80 mm which thus allowed a calculation of results up to approximately 1.25 kHz.

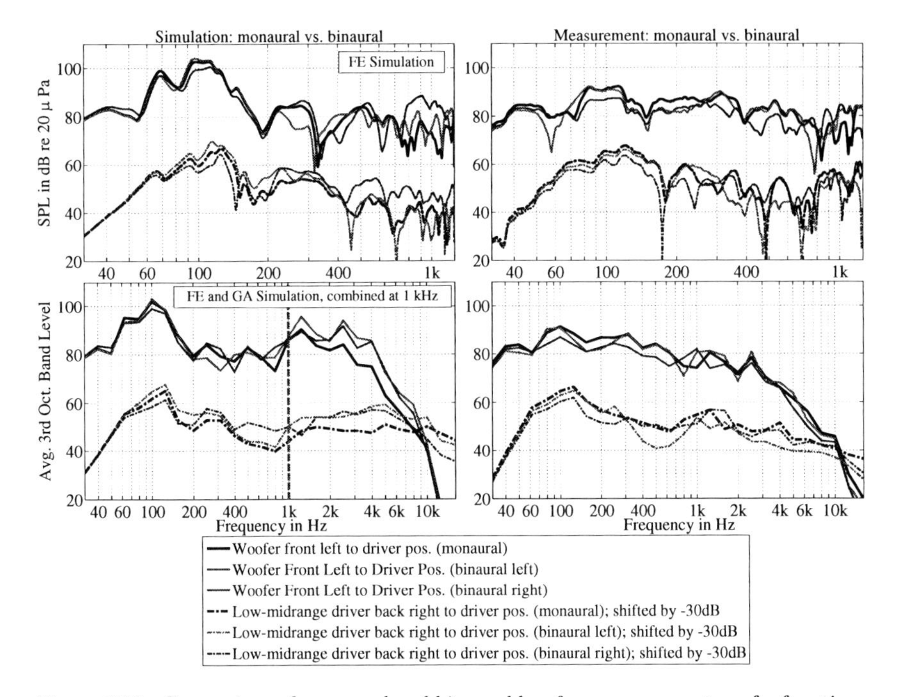

Figure 9.22.: Comparison of monaural and binaural low frequency room transfer functions and averaged band energies in simulation and measurement for car model 1 with 2 different car loudspeakers as sound sources.
The results are given for the driver's position. The microphone was positioned approximately at the driver's head position. The used loudspeakers are the woofer in the front left door and the low- to midrange driver in the back right door.

Figure 9.22 (a) gives a quantitative assessment of the low frequency impact of the dummy head, by comparing the FE simulation results with and without the head in the driver's position. Figure 9.22 (b) shows the according comparison for the measured transfer functions.

With regard to the FE frequency range, it can be seen that the order of magnitude of the differences between monaural and binaural results is very similar in measurement and simulation and that in both cases noticeable deviations (between monaural and binaural) can already be observed for frequencies above 200 Hz. It can thus be concluded that as expected the inclusion of a head model into the FE mesh well captures the binaural effects caused by diffraction at the dummy head and shoulders. However, taking into account the considerable additional modeling complexity of including the head model into the FE mesh and the overall quite high deviations between the measured and simulated results, it is doubtable if this enhanced receiver model also leads to a noticeable improvement in the subjective perception of the simulation results. The answer to this question is of course also closely related to the considered FE frequency range. While it is in a first approximation possibly admissible to neglect the head and shoulder model in the frequency range below 500 Hz, an FE simulation up to higher frequencies surely requires the head model to include the important binaural cues generated by diffraction effects at the head and shoulders in the mid-frequency range. Above 2 kHz even diffraction effects at the outer ear might have to be considered.

In the GA frequency range it is striking that the binaural simulations show a constantly higher level than the corresponding monaural results, while the measurements show very similar high frequency band energy levels in the monaural and binaural case. This difference is most likely due to the use of different dummy heads (with different head/ear geometries and microphones) in the measurement and the GA simulation. In particular, the set of HRTFs used in the simulation software *RAVEN* was obtained using the *ITA* artificial head (cf. photograph in sec. 6.2). Due to the largely different head and ear geometries of the ITA head and the *HEAD Acoustics* head it is clear that their corresponding HRTF data also differs considerably at least in the frequency range above $2-3$ kHz where the HRTF is influenced by the outer ear and pinna geometry. Moreover, the use of the free field correction which was applied to the HRTF data used in *RAVEN* (cf. section 6.2) also increases the high frequency band levels in the simulations. A similar correction of the microphone response could however not be conducted for the used *HEAD Acoustics* artificial head, which was supplied by our industry partner for the measurements. Finally, it should be mentioned that the absence of the head and shoulder geometry in the GA simulations might be problematic with regard to possible reflections at the head and torso. This question becomes all the more important when considering the influence of one or more occupants (with their full body) on the room transfer function inside the car compartment. Section 9.6 therefore addresses these questions on the basis of comparisons of measurements with different occupancies in the car compartment.

Figure 9.23 finally shows a direct comparison of measured and simulated room transfer functions for the left and right ear using the same source loudspeakers as before. The plots also reveal the already discussed large discrepancy between the binaural simulations and measurements in the high frequency range. In the following subsection we therefore report binaural simulation and measurement results for car model 2 which are both based on the *ITA* artificial head. As will be seen in the following section the use of the same dummy head with a fully calibrated input and output chain leads to a strongly improved match between simulated and measured high frequency levels and thus corroborates our implications that the observed differences for car model 1 can indeed mostly be attributed to the use of different dummy heads and different microphone equalization functions (in case of the HEAD acoustics head actually no equalization was used).

It is also interesting to note that in the case of the loudspeaker in the right back door the GA simulations unexpectedly generate a higher sound pressure level at the left than at the right ear, while the measurements show the expected opposite behaviour (i.e. higher levels at the right ear than at the left ear). By closer inspection of the GA results this error can most likely be attributed to the absence of a direct sound path and the negligence of diffraction effects at the car seats. With the absence of a direct sound contribution it is clear that the sum of the reflected contributions from different angles of incidence lead to rather random level differences between the left and right ear. Indeed, it was confirmed by listening to binaural auralizations based on the measured and simulated room transfer

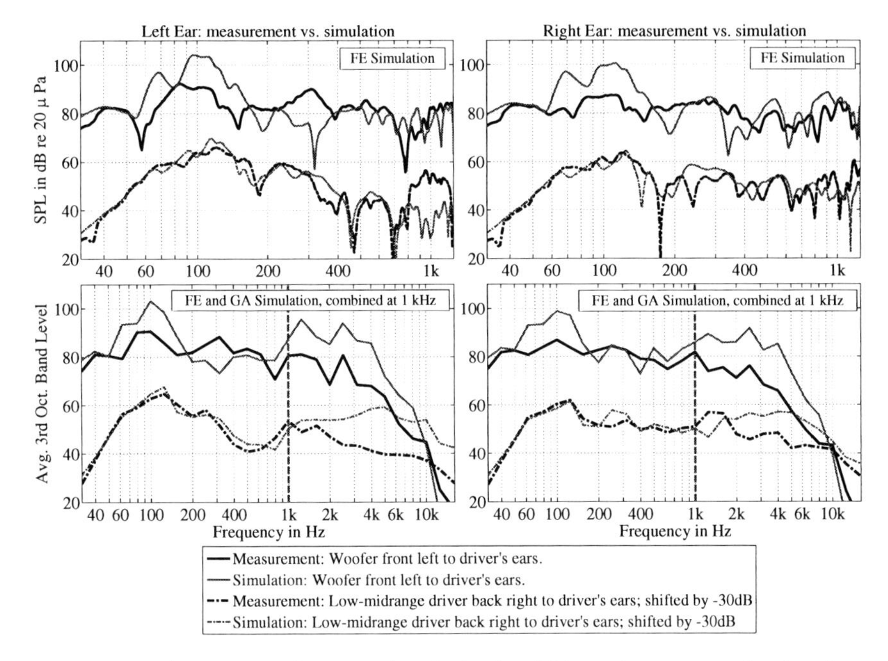

Figure 9.23.: Comparison of measured and simulated binaural room transfer functions and averaged band energies for the left and right ear in car model 1 with 2 different car loudspeakers as sound sources.
The results are given for the driver's position. The microphone was positioned approximately at the driver's head position. The used loudspeakers are the woofer in the front left door and the low- to midrange driver in the back right door.

functions, that this error leads to localization difficulties in the binaural GA simulations of this loudspeaker position. In contrast to this the localization of auralization based on the corresponding lowpass-filtered FE impulse responses ($f_c = 1\,\text{kHz}$) which include the diffraction effects appeared to be much better.

9.4.4. Monaural and binaural results with car loudspeakers (car model 2)

In order to found our analysis and conclusions on a wider basis of results, we have conducted an analogous measurement and simulation study also for car model 2. While the simulation settings and measurement setups were mostly the same as for car model 1, some important aspects need to be pointed out.

1. The acoustic reflection characteristics of the car materials were mostly described with the same boundary conditions as used for car model 1. Only for the car seats different boundary conditions were applied, since in contrast to the seats in model 1, which had a perforated leather cover, the seats in model 2 had a closed leather cover (the differences in the reflection characteristics of seats with perforated and non perforated leather cover are shown in figure 9.7). Despite the other differences in the model type and equipment version between car model 1 and 2, the remaining boundary materials appeared sufficiently similar (at least by visual inspection), to justify the use of the data obtained for car model 1. Moreover, it has to be taken into account that the existing differences in the acoustic reflection characteristics of the materials used in car model 1 and 2 would possibly be blurred by the considerable uncertainty in the determination of the boundary data.
2. Since different loudspeakers were used in car model 2 than in car model 1, all loudspeakers have been individually measured using the same setup as for the loudspeakers in car model 1 (see section 9.3).
3. The monaural results were conducted again using KE4 capsules positioned at head height on all seats. The binaural measurements were this time conducted using the *ITA* artificial head for better comparability with the *RAVEN* results.
4. For the sake of simplicity and to reduce the modeling complexity the FEM simulations were only carried out for an empty model of the car compartment without dummy head. The errors resulting from this simplification in the case of the binaural simulations are discussed at length in the previous subsection for car model 1 and shall not be repeated here.
5. All FE simulations were run using a mesh with an average element size of 80 mm. The simulations were therefore run up to a maximum frequency of 1.25 kHz.
6. As can be seen from figure 9.1 the GA model for car model 2 was modeled with much more detail than the GA model used for car model 1.
7. All measurements and simulations were fully calibrated to give absolute sound pressure levels in Pa per Volt at the loudspeaker inputs, which allows a direct comparison of the measured and simulated results without any further adjustment.

The results of the monaural and binaural simulations are given in figures 9.24 and 9.25 for the midrange driver in the front door on the driver's side and the midrange driver in the back door on the passenger's side. While all considerations made for car model 1 regarding the uncertainty of the input data obviously still hold for car model 2, the overall agreement of the results appears to be generally a little better than for car model 1. This is especially noticeable with regard to the agreement of the binaural high frequency energy levels between measurement and GA simulation. Thus the use of exactly the same dummy head and microphone equalization apparently helps to considerably improve the match in the high frequency energy levels. Further aspects related to the improved simulation quality in the FE and GA domain may be attributed to the fact that the midrange drivers in car model 2 are mounted in the upper part of the door liner and thus radiate the sound more freely into the passenger compartment. In contrast to this, the door loudspeakers of car model 1 were mounted in the bottom part of the door liners and thus radiated

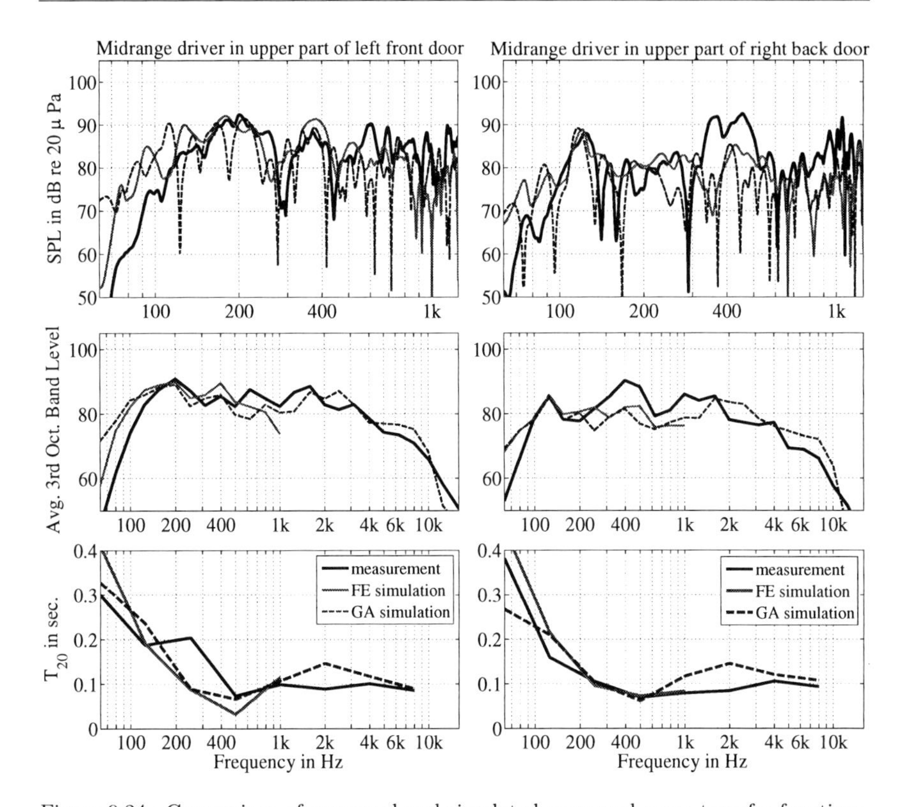

Figure 9.24.: Comparison of measured and simulated monaural room transfer functions, averaged band energies and reverberation times for car model 2 with two different car loudspeakers as sound sources.
The results are given for the driver's position. The microphone was positioned approximately at the driver's head position. The used loudspeakers are the midrange driver in the left front door and the midrange driver in the right back door.

the sound primarily into the footspace, which made the simulation results very sensitive to errors in the difficult to determine boundary definition of the car mats and the floor carpet. Moreover, the piston range of the used midrange driver with a diameter of 8.5 cm extends up to ≈ 1.2 kHz, which make the use of a piston source model with homogeneous velocity reasonable for the whole considered FE frequency range. Measurements of the membrane velocity at different positions on the loudspeaker membrane have also confirmed that considerable spatial variations of the membrane velocity only occur above roughly 1.3 kHz.

It should finally be mentioned, that we generally do not expect the increased detail in the GA model to significantly contribute to the improved simulation results. This is due to the fact, that at least for the lower and mid-frequencies a specular reflection at a very

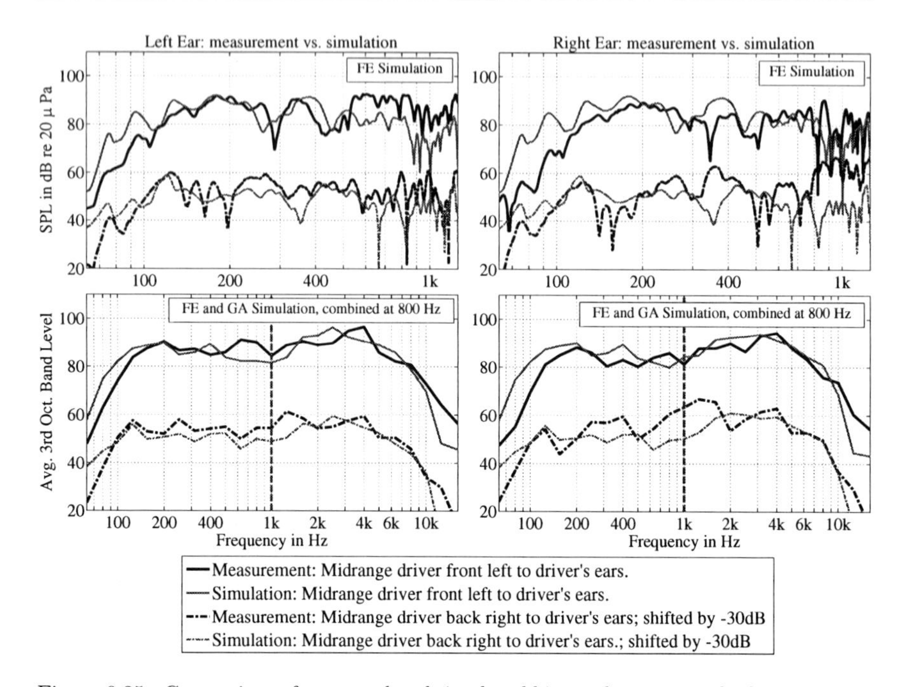

Figure 9.25.: Comparison of measured and simulated binaural room transfer functions and averaged band energies for car model 2 with two different car loudspeakers as sound sources.
The results are given for the driver's position. The used loudspeakers are the midrange driver in the left front door and the midrange driver in the right back door. The FE results are taken from the corresponding monaural simulations, since no dummy head was included in the mesh for car model 2.

small surface of a highly detailed region in the GA model does not necessarily constitute a more accurate reflection model than a reflection at a simplified larger surface. An improvement of the spatial distribution of reflection angles in the GA domain could only be achieved by assigning realistic scattering coefficients to these detailed regions in the GA domain. However, as already mentioned earlier the determination of this scattering data (by measurement or simulation) is extremely difficult or in some cases even impractical.

Summing up, it can be concluded that the results obtained for car model 2 clearly point out the potential of the combined approach to realistically predict the energy distribution of the sound field in the car passenger compartment throughout the whole audible frequency range. However, there still remain some open questions with regard to the applicable frequency range of the GA simulation, since depending on the source and receiver positions in the car passenger compartment, diffraction effects appear to play an important role even up to the upper mid-frequency range. However, an extension of the FE results to higher frequencies necessitates more complex source models and makes the inclusion of a dummy head model inevitable, to include the binaural cues in this frequency range.

9.5. Preliminary subjective assessment of results

For a preliminary subjective assessment of the simulation quality, the measured and simulated room transfer functions obtained for car model 1 and 2 were convolved with the same audio material which was used for the subjective evaluation of the recording studio simulations (excerpts from pop, hip hop and classical music as well as speech) and played back to a small number of selected, trained listeners via headphones. In addition to the evaluation of the listeners' comments we tried to conduct a similar comparison of the specific loudness characteristics of the measured and simulated auralization files as was done in the study of the recording studio. Unfortunately, due to the much bigger differences between measurements and simulations in the car compartment compared to the recording study, this analysis did not yield much valuable insights, despite the fact that it confirmed the considerable differences between measured and simulation files, which were already observed in the direct comparison of the room transfer functions. The comparative loudness plots are therefore not reported in the course of this thesis. The preliminary listening tests were carried out with monaural and binaural auralization files and the following conclusions can be drawn from the discussions with the listeners.

- No artefacts were reported for the monaural or binaural auralizations based on the combined simulation approach.
- As already expected from the differences in the measured and simulated room transfer functions (monaural and binaural) for car model 1, the subjective comparison confirms that considerable differences in the loudness and sound coloration can be heard between the auralizations based on the measured and simulated room transfer functions. With regard to the sound coloration it can be further stated that the differences between measured and simulated auralizations are much bigger than the differences between the pure GA and the combined FE-GA simulations. Thus a benefit of the low frequency FE extension is hardly heard in the auralizations of car model 1.
- Especially the simulations of the loudspeakers in the footspace of car model 1 appear to suffer from considerable errors with regard to the energy distribution throughout the whole audible frequency range. As already discussed earlier, this is presumably both due to errors in the source representation of the woofer loudspeakers (mounting, orientation, frequency characteristics) as well as to errors in the boundary characterization of the floor carpet and carmats in the footspace.
- The simulations of car model 2 show an overall improved match with the corresponding measurement files with regard to the loudness and sound coloration aspects and in a blind test it is hardly possible to judge which one of the auralizations is based on the measured and which is based on the simulated room transfer functions. However, in a direct A/B comparison differences between the measurements and simulations are still easily heard. A benefit of the FE extension with regard to an improved match in the bass response is still hardly audible.
- In the case of an obstructed direct sound path the localization quality of some of the binaural GA simulations considerably degrades. This is however generally not the case in the corresponding auralizations based on the measured binaural impulse responses. Thus, the poor localization quality in the GA simulations is most likely

caused by the negligence of diffraction effects at the seats. The strongest sound contributions in such a GA simulation may erroneously be generated by a random sound reflection at one of the almost rigid window surfaces.

- The localization quality was also assessed for the binaural FE simulations with dummy head. Despite the bandwidth limitation of these simulations to frequencies below 1.25 kHz it could be verified that in the case of an obstructed direct sound path the localization quality of the binaural FE simulations (lowpass filtered at 1 kHz) was considerably improved compared to the geometrical acoustics simulation.

  On the other hand it was found that the combined binaural simulations for car model 2, which were crossfaded at 1 kHz and which used FE results calculated without a dummy head model, showed a decreased localization quality, due to the lack of the binaural cues below 1 kHz. It can thus be concluded that the inclusion of a dummy head in the FE simulation is indeed very important already below 1 kHz. Moreover, since the localization problems in the GA domain, due to the missing diffraction effects, call for an extension of the FE calculations to higher frequencies, the inclusion of a dummy head model and even a rudimental ear geometry becomes all the more inevitable to realistically capture the binaural cues in the lower and mid-frequency range.

9.6. Additional measurements in the car compartments

In addition to the measurements of the RTFs for comparison with our simulation results, we conducted further measurements to investigate the influence of different factors on the sound field in the car compartment. These factors will be briefly discussed in the following subsections.

9.6.1. Influence of different occupancies

Due to the comparatively small volume of the car passenger compartment it is interesting to ask the question if the presence of one or more occupants inside the car passenger compartment considerably influences the sound field inside the car. This question was investigated by conducting measurements with the artificial head on the driver's position and a variable number of additional occupants (real persons) on the remaining seats in the car compartment. The measurements were conducted consecutively for all loudspeakers in the car compartment of car model 1. Figure 9.26 shows the results of these measurements using (a) only the loudspeaker in the driver's door, (b) only the loudspeakers in the back right door and (c) all loudspeakers in the car compartment at once. In order to mix the RTFs obtained for each loudspeaker, the RTFs were first individually bandpassed according to their specific frequency range of operation (the specifications were given by our partner in the automotive industry) and then added up to superimpose the loudspeaker contributions. Since all RTFs were measured re 1 V at the loudspeaker voice coils, this gives a reasonable weighting of all individual loudspeaker contributions. Although the applied up-mixing algorithms in real car audio systems are of course very different and possibly more complex, it is emphasized that this does surely not influence the general implications of the presented results.

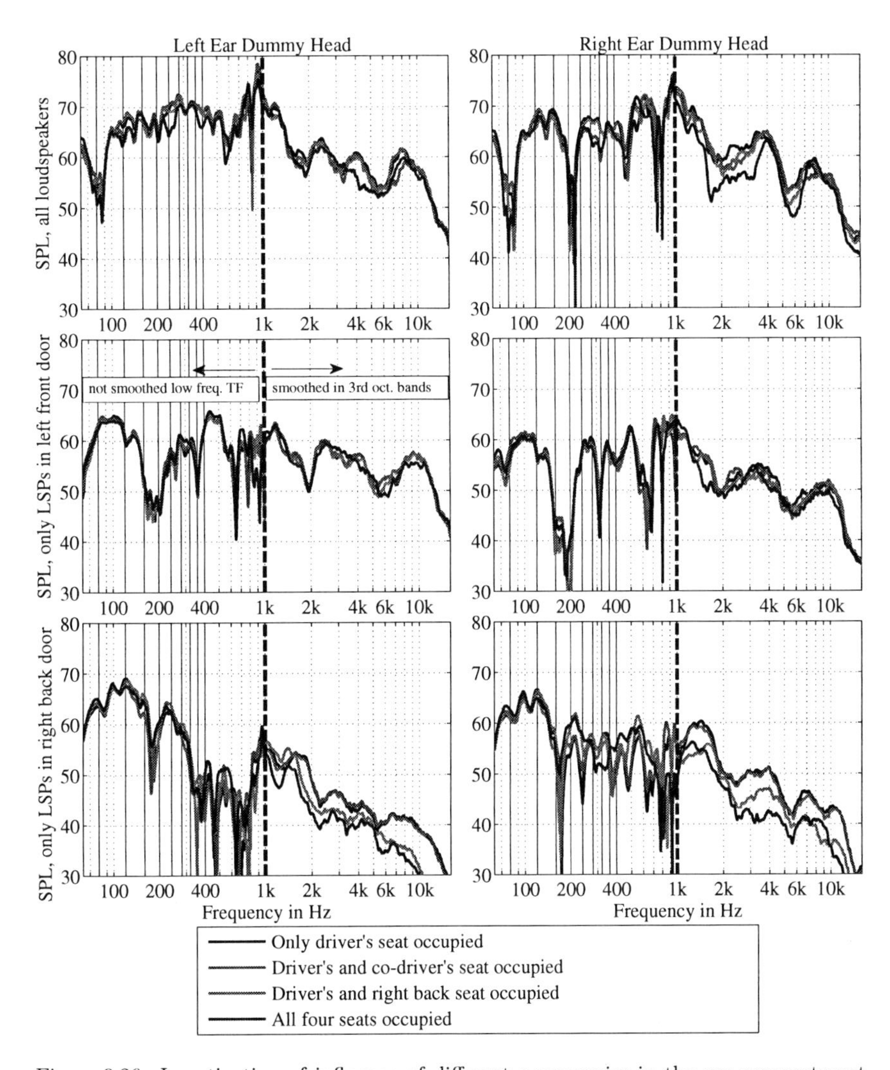

Figure 9.26.: Investigation of influence of different occupancies in the car compartment on the sound field at the driver's ears.
The binaural transfer function have been smoothed in 3rd octave bands for frequencies above 1 kHz for better distinction of the curves.

By looking at the results obtained in case (a) it becomes clear that due to the strong contribution of the unobstructed direct sound, the sound field at the driver's ear is hardly influenced by the presence of other occupants in the car. On the other hand, case (b) reveals that the sound field produced by loudspeakers in remote position's to the listener on the driver's seat is considerably influenced by occupants in close vicinity to these loudspeakers. In particular, the occupant sitting in the back right seat clearly obstructs some of the sound radiated by the loudspeakers in the back right door on its way to the driver's ear. However, it appears that with all loudspeakers playing (case (c)) the sound field at the dummy head's ears is mostly dominated by the loudspeakers closest to the dummy head and since this sound field is only marginally affected by the occupants on the other seats, the total influence of the occupancy in the car on the sound energy at the listener's ears is surprisingly small. Moreover, with all loudspeakers playing even the modal structure of the low frequency transfer function remains largely stable irrespective of the occupancy in the car compartment.

9.6.2. Influence of exact seating position in the car compartment

In order to assess the uncertainty in our simulation results caused by a possible maladjustment of the car seats and the dummy head, we have carried out another measurement set in the unoccupied car with the artificial head on the driver's seat for four different settings of the seat position. Except for the seat position (and thus also the position of the dummy head) everything else remained unchanged between the measurements. The measurements were carried out in car model 1 and the sound field was excited using sweep signals which were played back through the aux input of the car sound system. Thus, in contrast to the results of the previous subsection, in this case the mixing algorithm of the car audio system was used. The results of the measurements are shown in figure 9.27, which also gives detailed information on the adjustment of the seat position. Although the EQ settings of the car audio systems were set completely flat, a strong bass boost of approximately 20 dB is observed for frequencies below 100 Hz, which is due to the inherent tuning of the car sound system. A direct comparison of the different curves reveals that considerable differences between the obtained room transfer functions can partly already be observed from 200 Hz on, depending on the setting of the seat position. While it is generally not possible to deduce implications on the subjective importance of these differences from the direct comparison of the RTFs, it is still important to note that the induced changes in the measurement setup introduce noticeable variations in the measurements already in the bass and lower mid-frequency range.

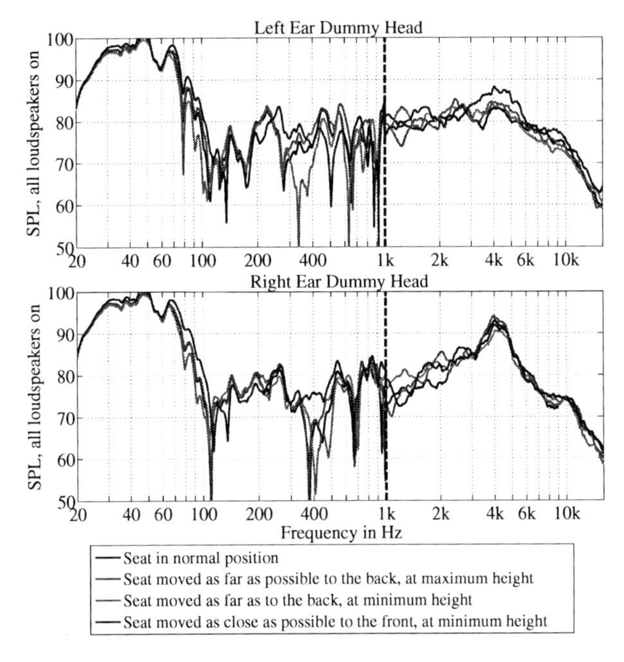

Figure 9.27.: Investigation of influence of different seat positions in the car compartment on the sound field at the driver's ears.
The binaural transfer function have been smoothed in 3rd octave bands for frequencies above 1 kHz for better distinction of the curves.

9.6.3. Influence of different listeners

In a further measurement set, we have investigated the impact of different listeners on the sound field in the car compartment. We have therefore conducted measurements with 4 test persons wearing ear plugs with miniature microphones in it. The plugs were fitted in the test persons' ear canals, such that the microphones were placed as good as possible at the blocked ear canal entrance. All test persons were seated on the driver's seat for a fixed seat adjustment and in a similar sitting posture. The differences in the binaural room transfer functions are thus only caused by the different body, head and ear shapes of the listeners. The measurements were carried out in car model 1 and the sound field in the car was again excited using sweep signals which were played back through the aux input of the car sound system with the strong bass boost below 100 Hz. It can be seen that the listener's body, head and ear shape as well as the exact position of the microphones only have a rather small influence on the sound field below 500 Hz. In contrast to that, the smoothed room transfer functions for frequencies higher than 1 kHz show differences of up to ± 5 dB, which might however be partially due to differences in the exact positioning of the microphones in the ear canals.

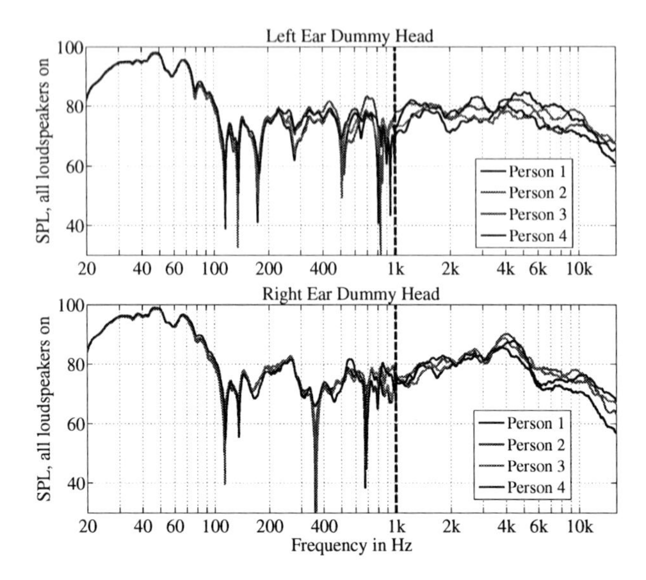

Figure 9.28.: Investigation of influence of different listeners on the sound field at the driver's ears.
The binaural transfer function have been smoothed in 3rd octave bands for frequencies above 1 kHz for better distinction of the curves.

9.6.4. Influence of the occupant's clothing on the low frequency transfer function

Finally, we have investigated the influence of different clothing of the occupants on the binaural transfer function in the car compartment. For the measurements the ITA artificial head was differently "dressed" (see figure 9.29) and placed on the driver's seat. The measurements were carried out in car model 1 and the sound field was excited again using sweep signals which were fed into the aux input of the car sound system. The results of the measurements for the left and right ear are shown in figure 9.30. It can be seen that the clothing does only marginally influence the sound field in the bass and lower-mid frequency range. Moreover, the average band levels for the higher frequencies also remain almost unchanged irrespective of the clothing. An additional subjective assessment of the differences, which was conducted by convolving the measured binaural transfer functions with different audio stimuli, has further confirmed that at least from a perceptual point of view the different clothing does, if at all, only marginally affect the sound field at the listeners ears.

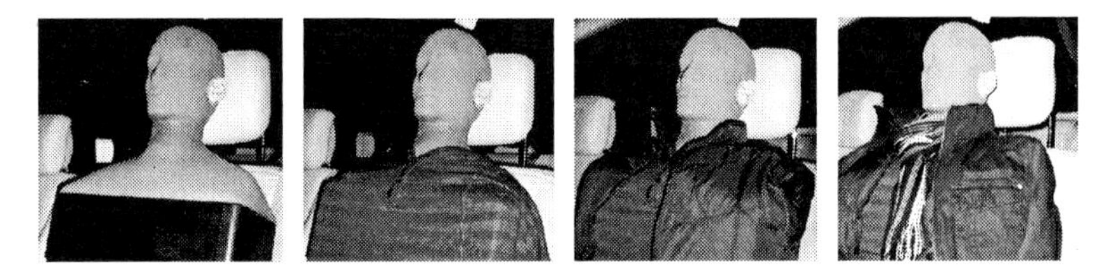

Figure 9.29.: Dummy head with different clothing.

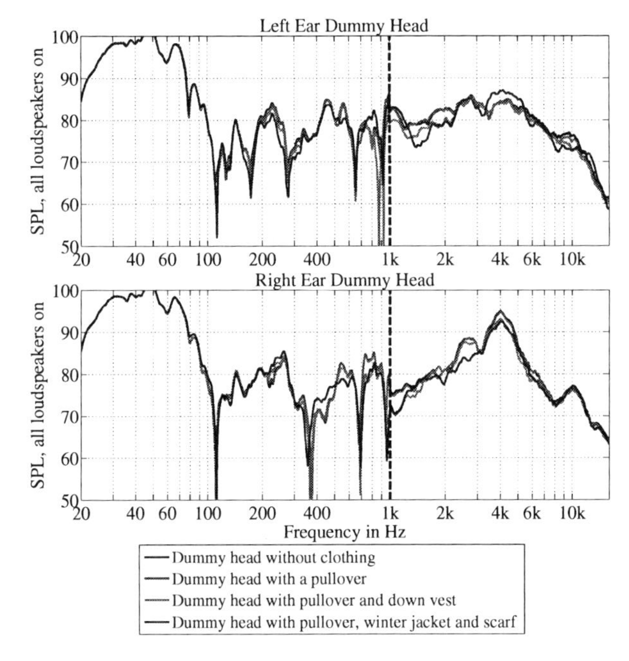

Figure 9.30.: Investigation of influence of different clothing on the sound field at the driver's ears.
The binaural transfer function have been smoothed in 3rd octave bands for frequencies above 1 kHz for better distinction of the curves.

9.7. Summary

The present study gives a detailed elaboration on the challenges and limitations of sound field simulations in a car passenger compartment and discusses the potential of a wave-based low frequency FE extension of the classical geometrical acoustics simulation approach. The investigations have shown that a classical GA based simulation approach faces various problems in the car passenger compartment, which are related to different wave and near field effects which can in general not be captured by an energy based sound ray model. While it was a-priori expected that the GA simulation approach would fail to produce realistic results in the modally dominated low frequency part of the room transfer function, it was found that the problems of the GA approach extend to considerably higher frequency ranges than indicated by the Schroeder frequency ($\approx 400\,\text{Hz}$ in the car compartment), which is often referred to as the lower frequency bound for the application of geometrical acoustics. These limitations are considered to be mostly due to strong diffraction effects at the car seats. Taking into account the size of the backrests and headrests the influence of these diffraction effects surely extends to the mid-frequency range. Moreover, the close proximity of sound sources and receivers to the hard reflective windows raises further questions and doubts regarding the applicability of geometrical acoustics in the lower and mid-frequency range.

Since all the above mentioned wave effects are in general fully captured by the wave-based FE approach, the substitution of the lower and mid-frequency part of the GA simulation with a corresponding FE simulation appears very reasonable and is nowadays also feasible with regard to its computational complexity; e.g. the simulation of the car passenger compartment in the frequency range from 10 Hz to 3 kHz lasted about 2 days on a conventional personal computer using the software *Virtual Lab*. Unfortunately, the application of the FE method to the simulation of the sound field in the car compartment introduces new challenges regarding the source, receiver and boundary representation and it has been shown in the course of this study that improved simulation results can only be achieved if the used simulation input data is of very high accuracy. Since this is extremely difficult to achieve due to the high complexity of the boundary materials and sound source configurations in the car compartment, a satisfactory improvement in the prediction of the low frequency transfer function could so far not be achieved (at least not for all source and receiver pairs). Nonetheless it has to be mentioned, that the consideration of the diffraction effects in the FE simulations already leads to an improved localization quality in the binaural simulations (especially in the case of an obstructed direct sound path), which confirms that the FE approach does indeed constitute the more realistic sound field model.

While it is generally difficult to determine which influencing factors most strongly contribute to the overall simulation error, the following summary gives some guidelines for the sound field simulation in car passenger compartments, which are deduced from the broad range of simulation and measurement results that were analyzed in the course of this study:

- For high quality binaural simulation results the inclusion of a dummy head model in the FE model is strongly advised. Depending on the upper frequency limit of the FE simulations a rudimentary model of the outer ears should also be included in the model.
- In order to realistically capture all relevant diffraction effects in the car passenger compartment, the FE simulations should be run up to at least $3-4\,\text{kHz}$. However, this necessitates a refined source model (see below).

- The geometrical acoustics simulation captures the physics of the high frequency part of the wave propagation sufficiently well, where diffraction does not play a dominant role.
- A realistic representation of the modal low frequency characteristics of the sound field can only be achieved with high quality complex impedance data for all boundary surfaces in the car compartment. The challenges in conjunction with the determination of this data, are a crucial point with regard to the overall simulation quality. In particular, the often encountered use of real valued impedance data, which is obtained from measured absorption values, generally leads to a considerable error in the modal characteristics of the sound field.
- An extension of the FE frequency range to frequencies far above the piston range of the considered source loudspeakers necessitates a refinement of the FE sound source model with regard to the consideration of the spatial velocity distribution on the membrane. Moreover, it is important that the source surfaces (loudspeaker membranes) are modeled with their actual size, position and mounting conditions in the car compartment, since this affects the directivity pattern of the source (e.g. mounting in a recess, tilting of the loudspeakers within the recess).
- In the GA domain, it is also important to consider the exact position and orientation of the sound sources, since due to the strong focussing of the radiation pattern at high frequencies and the short source receiver distances, small positioning and orientation errors may especially affect the direct sound contribution at the receiver position.
- Finally, it is important to mention, that even in the case of very accurate boundary and source data for the FE simulations a precise prediction of the modal structure of the sound field remains impossible for frequencies far above the Schroeder frequency ($\approx 400\,\mathrm{Hz}$). This statement is based on the fundamental findings of Schroeder [1954b], who showed that in the frequency range with high modal overlap already small variations in the room geometry lead to a completely stochastic reformation of the room eigenmodes. Considering the different possible occupancies and adjustable seat positions in a car compartment, the goal of a room acoustic simulation for frequencies far above the Schroeder frequency can therefore only be the best possible prediction of the sound decay and sound energy in all frequency bands.

  Consequently, the aim of an extension of the FE simulations to the upper mid-frequency range, is not a detailed prediction of the fine structure of the room transfer function, but a more realistic representation of the time and frequency dependent energy distribution in the car compartment, compared to a pure geometrical acoustics based simulation. The potential of the FE simulations in this context lies in the full consideration of all near-field and diffraction effects in the car compartment. Unfortunately, our simulation results reveal that the quality of the used boundary and source data is yet not sufficient to fully exploit the benefits of the improved sound field model in the FE simulations.

10. Conclusion

10.1. Summary

The present thesis gives a comprehensive insight into the challenges and limitations of sound field simulations in small rooms including all aspects of sound generation, sound reflection and sound reception at the receiver. Special focus is put on the possible benefit of a low frequency extension of existing hybrid geometrical acoustics tools by means of room acoustic finite element simulations. Taking into account that the FEM gives an approximate solution of the Helmholtz wave equation in a bounded domain including all wave-effects such as edge diffraction, interference and standing waves, this powerful combination enables the realistic simulation of sound fields even in small rooms, where the modally dominated part of the room transfer function may extend far into the audible frequency range. However, despite the fact that the FE method is a well-established and highly developed tool in engineering sciences, its application to room acoustic problems introduces far-reaching and yet unresolved questions regarding the source, boundary and receiver representation in the FE simulations. In particular, the boundary and source representations in the FE domain differ considerably from the respective representations in the GA domain and it is shown in the course of this thesis that a straight-forward conversion of existing geometrical acoustics input data to the FE domain is generally not admissible if high requirements on the simulation accuracy exist.

The present thesis therefore introduces a complete framework for the combination of wave- and ray-based room acoustic simulation results and gives detailed guidelines for the best-possible determination of all necessary input data for both simulation domains. Thus, the first part of the thesis summarizes the fundamentals of the applied simulation methods and discusses the theoretical links between wave- and ray-based room acoustics. Moreover, the thesis gives an overview of existing methods for the determination of the boundary, source and receiver data in both simulation domains and discusses their respective strength and weaknesses, where special focus is put on possible improvements of existing methods for the determination of the boundary and source data in the FE domain.

Additionally, the thesis discusses selected isolated aspects which are related to the realistic simulation of sound fields in small rooms. These topics include a study on the potential of the image source method to predict the low frequency characteristics of a room transfer function, a study on the efficient modeling of porous absorbers in the FE domain and finally a study on the possible low frequency coupling of the excitation velocity of a loudspeaker source to the sound field at the loudspeaker membrane. A summary of the results of these studies is given in the respective chapters and shall not be repeated here.

The core of the thesis is given by an extensive comparison of measurement and simulation results for three types of acoustically relevant small spaces (a scale-model reverberation chamber, a recording studio control room and two different car passenger compartments). For each room considerable efforts have been made to obtain a best-possible a-priori assessment of all necessary material and source data for the simulations. While the results of the presented application studies generally confirm the potential of the combined approach to realistically simulate room acoustic sound fields in the whole audible frequency range, it is also shown that an actual objective and subjective benefit of the FE low frequency extension can generally only be achieved if highly accurate data exists for both the geometric room model as well as for the boundary, source and receiver conditions. However, especially with regard to the determination of the acoustic surface impedances at the room boundaries this does not always appear possible and certain inevitable inaccuracies have to be accepted.

In order to assess the influence of these inaccuracies in the simulation input data, a thorough comparison of the measured and simulated impulse responses was conducted both on the basis of objective and subjective evaluations of the results. With regard to the subjectively perceived quality of the simulation results it can be stated that the simulated monaural and binaural impulse responses showed no audible artefacts and that the overall energy distribution in time and frequency was generally well captured by the simulations in all considered rooms. However, the results obtained in the three very different rooms clearly show that the simulation accuracy considerably degrades with increasing complexity of the room geometry and boundary conditions. Moreover, it is important to mention that even with the FE method a precise prediction of the fine structure of the room transfer function appears impossible in the frequency range far above the Schroeder frequency. This is due to the fact that in this frequency range the strongly overlapping and highly dense room modes are affected by very small changes in the atmospheric conditions, the geometry as well as the boundary and source conditions which appears impossible to consider in a room acoustic simulation. Thus, possible fields of application of the FE extension in room acoustic simulations lie in the prediction of the modally dominated low frequency part of the room transfer function and in the prediction of sound fields that are strongly affected by near-field or diffraction effects as in the car passenger compartment. In such cases these wave-effects may considerably influence the energy distribution even at frequencies far above the Schroeder frequency and thus an extension of the FE frequency range to the stochastic part of the room transfer function appears reasonable although the fine structure of the RTF can of course not be accurately predicted.

Summing up, it can be concluded that despite the general potential of the low frequency FE extension to realistically predict the modal structure of the modally dominated part of a room transfer function, the Achilles' heel of room acoustic FE simulations appears to be the determination of realistic impedance conditions on the room boundaries. Consequently the application of the finite element method to room acoustic applications calls for improved measurement techniques for the acoustic surface impedance. This and other topics for future work are summarized in the following section.

10.2. Outlook

The present thesis covers a broad spectrum of topics and studies that are related to the realistic physically based simulation and auralization of sound fields in small rooms. However, naturally, not all of the questions and subjects that were raised in the course of this thesis could be satisfactorily resolved and thus interesting topics for future work emerge in many different areas related to this thesis. These topics can be roughly subdivided into four different areas:

1. The further development of the used FE and GA simulation algorithms as well as the presented combination method.
2. An improvement of existing and/or the development of new, more reliable measurement methods to determine the necessary boundary and source data for the simulations.
3. A better quantification of the influence of individual errors in the sound field models themselves or in the assigned input data on the resulting sound field in the room.
4. The development of improved objective and subjective evaluation techniques for the benchmark of the quality of room acoustic simulations in small rooms.

The following paragraph gives a more detailed description of possible work packages located in these four areas.

With regard to the first point it is clear that possible further improvements of the used algorithms and sound field models lie mainly in the geometrical acoustics domain, since the wave-based FE method already gives a very realistic model of the sound propagation in a closed cavity. Further improvements of the FE method would only be possible by considering a mutual coupling of the airborne sound propagation in the room with suitable sound propagation models for the boundary materials. Such a detailed description of each single boundary layer is however out of scope of common applications in room acoustics. Moreover, the considerable uncertainty in the determination of all necessary material parameters and mounting conditions further complicates the application of such models.

On the other hand, the most important improvement with regard to the geometrical acoustics domain surely appears to be the inclusion of edge diffraction models. As discussed in detail in section 4.2.3 such models are currently on the verge of becoming available in state-of-the-art geometrical acoustics simulation tools and first implementations and validations have already been published Schröder [2011]. Moreover, the results of section 7.1 suggest that in simple concave rooms the inclusion of angle dependent, complex reflection factors into the image source method leads to improved simulation results in the frequency range close to the Schroeder frequency with moderate modal overlap. While the significance of this result in the case of more complex rooms needs to be checked by further investigations, it should be mentioned that the inclusion of angle dependent, complex reflection factors into the image source calculation comes at a very low cost with regard to the necessary adjustments to the simulation algorithm. The bigger problem appears to be the determination of the complex and angle dependant reflection factor data. However, under the assumption of locally reacting boundaries this data can be extracted from the acoustic surface impedance of the boundary, which in the case of a combined

FE-GA simulation approach needs to be determined for the low frequency FE simulation anyway. Thus, the inclusion of complex, angle dependent reflection factors into the image source method may yield an improved quality of the early part of the room impulse response, at least if it is dominated by contributions from specular reflections.

With regard to the determination of the boundary and source conditions, the results presented in sections 8 and 9 clearly reveal the need for improved measurement or modeling methods. In particular, it has been shown that none of the presented methods for the determination of the acoustic surface impedance was able to provide reliable low frequency results for inhomogeneously layered and non-flat boundary materials. This problem is even more aggravated in the case of low absorbing materials. Despite the fact that some improvements and guidelines for existing impedance measurement techniques are suggested in sections 5 and 6, considerable future work needs to be invested into the development of new methods, that allow the determination of the acoustic surface impedance even in these very difficult cases. Current work at the *ITA* is therefore focussed on improvements of the Microflown in-situ measurement technique and first results are presented in the master's thesis by van Gemmeren [2011]. Moreover, it is recommended to build a new larger impedance tube according to the specifications and suggestions given in section 5.1. It should finally be mentioned, that so far the only way to obtain consistent broadband impedance and absorption data is the use of the presented two-port network model for layered absorbers. However, this approach surely reaches its limits in the case of very complex or inhomogeneous material structures like a car dashboard or door liner. For these materials a reliable in-situ measurement method is desired. However, at the present state the investigated Microflown in-situ method cannot be applied to such materials for various reasons [van Gemmeren, 2011].

The simulation results of the car passenger compartment clearly show that certain errors in the simulation models or the simulation input data lead to considerable differences between the measurement and simulation results obtained in this space. While it is very likely that the overall simulation error is caused by a number of different influencing factors, it is nevertheless important to investigate in how far certain factors contribute quantitatively to this overall error. In particular, further investigations should focus on the identification and quantification of the most crucial factors deteriorating the simulation quality. The presented study in the scale model room shows that in the case of a well controlled measurement and simulation scenario such an isolation of individual influencing factors on the simulation accuracy is indeed possible. On the other hand, such a separation of errors appears almost impossible in the case of much more complex rooms, like the recording studio or the car passenger compartment. Recent work of the author and co-workers at *ITA* of RWTH Aachen University has therefore focused on the setup of a controlled variable measurement environment in order to benchmark the quality of room acoustic simulation software based on the true measurement result in the considered room (both regarding the used algorithms as well as the simulation input data). First results of these efforts have been presented by Pelzer et al. [2011a]. However, while this approach allows to isolate certain influencing factors in room acoustic simulations and assess their impact on the room acoustic sound field, it does not by itself give an indication of the perceptual relevance of the observed differences.

Thus, with regard to the quantification of the simulation quality, a major challenge for future work lies in the definition of a comprehensive framework for the perceptually-based comparison of room sound fields. Such a framework should include suitable listening tests for the evaluation of the similarity of measurement- and simulation-based auralizations as well as the definition of perceptually-motivated objective measures, which can be extracted from the room impulse response. Regarding the former a first step in this direction was made by Aretz and Jauer [2010], who propose a terminology and suitable listening tests for the comparison of measured and simulated auralizations. In particular it is suggested that "the quality of a room acoustic simulation shall be evaluated on the basis of its plausibility, its realism with respect to the listeners expectations and its similarity to the according measured auditory events" [Aretz and Jauer, 2010]. In the context of the definition of suitable objective measures, it is important to mention that an objective evaluation based on a direct comparison of the time and frequency characteristics of the room transfer function (spectrogram, transfer function, impulse response, energy decay curve) only gives an incomplete picture of the perceptually relevant factors for the simulation quality. This is especially true in the case of binaural simulations, where important influencing factors on the simulation quality such as the sound localization can hardly be evaluated by such means. Interesting work in this field was conducted by Lee and Cabrera [2009], Lee et al. [2011] who suggest an evaluation of room impulse responses based on the so-called loudness decay function, which is intended to be more closely related to the sound experienced by the listeners than typically used room acoustic parameters (such as given in the ISO 3382-1:2009 standard). However, in the case of the very short impulse responses as considered in the course of the present thesis it is questionable if this approach can be successfully applied. Another interesting approach presented by van Dorp Schuitman and De Vries [2008] consists of the use of auditory models to extract information on the room acoustics in a considered space. Similar to the intention by Lee this work also aims at the definition of objective measures that are closely related to human auditory perception.

It should finally be emphasized that the already very high quality of state-of-the-art room acoustic simulation tools on the one hand and the extreme complexity of the fine structure of room acoustic sound fields on the other hand, make further improvements in the simulation of room acoustic sound fields very difficult to achieve. In particular, it is striking that in a direct A/B comparison subtle differences between the measured and simulated sound fields appear to be always audible even for the simplest and most well-controlled rooms (such as the reverberation chamber used in Pelzer et al. [2011a]). On the other hand, even for the most complex rooms, such as the car passenger compartment, listeners were hardly able to tell, which one of the presented sound stimuli is based on the measured and which one is based on the simulated impulse response. Even in the case of the purely GA-based simulations the plausibility of the simulation results remained basically unchanged.

Thus, reasonable applications for the presented combined simulation approach are possibly constrained to expert applications where the realistic recreation of the acoustics of an already existing or planned room is vital in order to make the right acoustic design decisions. However, while up to now the uncertainty in the simulation input data appears to be generally too high to fully rely on the absolute simulation results, the investigation of relative changes in the sound field which are induced by changes in the room design appears already reasonable.

Acknowledgements

I would like to take the opportunity to express my deep gratitude and say "Thank you!" to my family, friends and colleagues who directly or indirectly contributed a great deal of inspiration, motivation, work and helpful discussions during the development process of this thesis.

First of all I would like to thank Prof. Dr. rer. nat. Michael Vorländer, head of the Institute of Technical Acoustics (ITA) of RWTH Aachen University, for the opportunity to work as a Ph.D. student at the ITA and to work and study independently and at my own pace throughout the whole time at the ITA. His supervision, encouragement and fruitful discussions throughout the whole Ph.D. thesis always helped to keep my research work focussed and on track. Moreover, I am grateful that he supported my job-accompanying business studies during my first two years at the institute. I would also like to thank Prof. Dr.-Ing. Otto von Estorff of the Institute of Modelling and Computation at the Hamburg University of Technology for taking on the role as the second examiner of my Ph.D. thesis.

A very warm thanks goes to all colleagues and students at the ITA for creating and sustaining an immensely positive, motivating and helpful atmosphere at all time and in every corner of the institute. Representative of all those inspirational people that I met and worked with at the institute, I want to highlight a few colleagues that contributed to my work in one or several special ways. In particular, I would like to thank Dr.-Ing. Gottfried K. Behler for literally always having an open door and ear for all kinds of scientific, professional or personal problems or questions that could possibly trouble his colleagues. Further thanks go to the staff at the electrical and mechanical workshop of the ITA, namely Rolf Kaldenbach, Hans-Jürgen Dilly, Uwe Schlömer and Thomas Schäfer, for their patience and determination when dealing with many different challenging tasks in the context of my Ph.D. thesis (and sometimes even outside my Ph.D. thesis. Their occasional emergency service on my guitar equipment two days before a gig was really invaluable). Many more colleagues, of which I can only name a few, contributed in various ways to the completion of my thesis and I am very grateful for that. As the supervisor of my diploma thesis Dr.-Ing. Andreas Franck introduced me to the world of finite element simulations and gave me the opportunity to learn from his vast expertise in this domain. Pascal Dietrich initiated the ITA Toolbox for Matlab which helped to bundle and standardize the programming efforts at the institute and considerably facilitated the measurement and acoustic signal processing of all measurement and simulation data used in my work. Martin Guski deserves special thanks for being the most helpful colleague I could possibly imagine. Moreover, I would like to thank my office colleagues Renzo Vitale and Roman Scharrer for being such a great and helpful company and for answering my countless questions (Roman would definitely become my telephone joker if I ever make it to a TV quiz show).

I am also very grateful for the excellent work that was contributed by all the diploma students and student workers that have worked with me over the past five years. Without their valuable contributions and results the research work in the present Ph.D. thesis would have surely not come that far. Namely I would like to mention Sebastian Schmidt, René Alexander Nöthen, Julian Knutzen, Ramona Bomhardt, Martin Praast, Lucas Jauer and Jan van Gemmeren. Moreover, I would like to express my gratitude to Josefa Oberem for thoroughly proofreading the entire thesis and for helping me out with some tough formatting issues.

Further thanks go to our project partners at Daimler AG, namely Norbert Niemczyk, Mario Fresner and Jochen Linkohr, for giving me the opportunity to conduct a great deal of my Ph.D. work in the course of a challenging and interesting industry project on sound fields in car passenger compartments. Without their on-going support during the whole project and the supply of test cars, CAD models, test materials and car loudspeakers many of the presented measurement and simulation results could have not been obtained.

Most importantly, I would like to thank my girlfriend Melanie, my family and my friends for their love and mental support throughout the whole time of this Ph.D. thesis and for putting up with my moods, when things sometimes did not work out as smooth as desired.

A. Appendix

A.1. Inclusion of the Thiele-Small loudspeaker source model in a room acoustic FE simulation.

The present section derives the coupling conditions for the inclusion of an active two-port network into a room acoustic FE simulation. In the context of room acoustic FE simulations the active two-port may be represented by the electro-acoustical network model of a loudspeaker source. The parameterization of such an electro-acoustical network model is given by the Thiele-Small parameters which are discussed in section 6.1.2.

The full inclusion of an electro-acoustical loudspeaker model into the FEM simulations allows (at least in the piston range of the considered loudspeaker) to use the input voltage at the loudspeaker voice-coils as the source parameter instead of the much more difficult to assess membrane velocity. Moreover, it is possible to account for mutual coupling between loudspeakers in multiple driver configurations. While it was shown in section 7.3 that this is generally negligible for typical room acoustic setups, useful applications might be found for the investigation of sound fields in very small closed cavities (calibrator cavities, loudspeaker enclosures) or for the investigation of subwoofer arrays. Further applications might also exist in the field of hydro or structure borne acoustics, where the feedback of the radiation impedance on the source velocity is generally much stronger.

In the FE fluid domain the loudspeaker is described as a radiating surface with normal velocity v_n. Using the conservation of momentum equation (eq.: 3.1 (a)) this excitation velocity can be defined as a Neumann boundary condition (cf. eq.: 3.14) for the Helmholtz equation. However, since the excitation velocity is dependent both on the input voltage at the loudspeaker voice coils as well as on the resulting sound pressure at the loudspeaker membrane, suitable coupling conditions need to be defined. The sought-after relations can be described by a two-port network given in admittance matrix representation (y-parameters):

$$\begin{bmatrix} I_{\text{inp}} \\ v_{\text{mem}} \end{bmatrix} = \begin{bmatrix} y_{11} & y_{12} \\ y_{21} & y_{22} \end{bmatrix} \cdot \begin{bmatrix} U_{\text{inp}} \\ p_{\text{mem}} \end{bmatrix} \tag{A.1}$$

where the signs of the input voltage and current (U_{inp}, I_{inp}) as well as of the membrane pressure and velocity (p_{mem}, v_{mem}) are defined as shown in figure A.1. By inserting the second equation of the matrix system A.1 into equation 3.1 (a) a mixed boundary condition at the radiating surface can now be given as follows:

$$-\frac{\partial p_{\text{mem}}}{\partial \boldsymbol{n}} - j\omega\rho_0\, y_{21}\, U_{\text{inp}} - j\omega\rho_0\, y_{22}\, p_{\text{mem}} = 0 \tag{A.2}$$

Inserting this into the integral form of the Helmholtz equation with boundary conditions (eq.: 4.1) and applying Green's first identity as described in section 4.1 yields the following integral equations for the fluid domain Ω and the radiating surface Γ_{mem}:

$$\int_{\Omega} -\nabla \bar{w} \left(\nabla p + k^2 p\right) \,\mathrm{d}\Omega - \oint_{\Gamma_{\text{mem}}} \bar{w}\, j\omega\rho_0 \left(y_{21}\, U_{\text{inp}} + y_{22}\, p_{\text{mem}}\right) \mathrm{d}\Gamma \tag{A.3}$$

where we have omitted all source terms in the fluid domain Ω and all possible Dirichlet, Neumann or Robin conditions on the other boundary surfaces of the fluid domain for better readability of the equation. Figure A.1 shows a schematic illustration of the resulting FE system of equations for an active two-port which is coupled to the FE fluid domain.

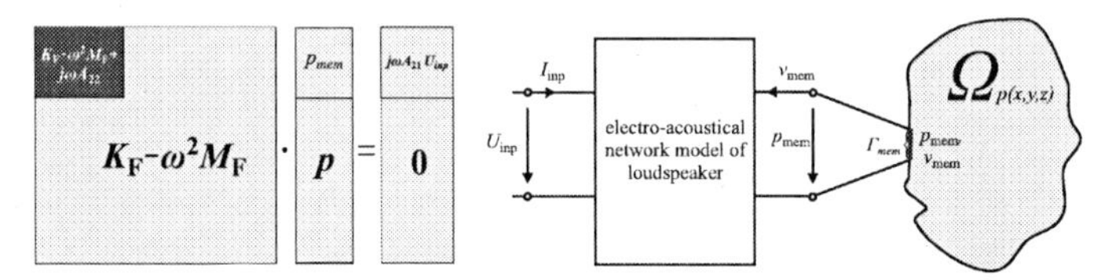

Figure A.1.: Schematic illustration of the FE system of equations for a loudspeaker source (represented by a electro-acoustical network model) which is coupled at its membrane surface to the adjacent fluid domain.
The coupling is modeled based on the admittance matrix parameters of a two-port network which is driven by the loudspeaker input voltage U_{inp}. The admittance matrix parameters can be deduced from the electro-acoustical network model of the considered loudspeaker source. The notation for the FE matrices is chosen according to equation 4.4. The indices of the damping matrices A_{ij} are chosen according to the indices of the corresponding admittance matrix terms y_{ij}. For simplicity all other walls in the fluid domain are considered rigid and no further sound sources are present within the fluid domain.

Thus the electroacoustical network model of the loudspeaker contributes an active source term ($y_{21}\, U_{\text{inp}}$) and a passive admittance term (y_{22}) to the integral form of the FE formulation. For a better understanding of the above equation it is interesting to recall the definition of the used admittance matrix parameters, which are:

$$y_{21} = \frac{v_{\text{mem}}}{U_{\text{inp}}}\Big|_{p_{\text{mem}}=0} \quad \text{and} \quad y_{22} = \frac{v_{\text{mem}}}{p_{\text{mem}}}\Big|_{U_{\text{inp}}=0} \tag{A.4}$$

Thus the term y_{22} describes the acoustical admittance of the loudspeaker membrane if the voice coil terminals are short-circuited. In this case the source term in the FE formulation vanishes since $U_{\text{inp}} = 0$ and the membrane surface is modeled as a fully passive impedance boundary condition with impedance $Z_S = \frac{1}{y_{22}}$. On the other hand the term $y_{21}\, U_{\text{inp}}$ corresponds to the excitation velocity if the loudspeaker is working on a zero radiation impedance ($p_{\text{mem}} = 0$), in which case however strictly speaking no sound power is induced into the fluid domain.

It is finally important to compare the fully coupled fluid-loudspeaker model to the source model suggested in section 6.1.2[1]. By direct comparison of the methods, it can be seen

[1] In section 6.1.2 it was suggested to determine the loudspeaker radiation impedance from an FE simulation with an idealized piston source with velocity $v_{\text{mem}} = 1\,\mathrm{m/s}$. This radiation impedance is then inserted into the electro-acoustical network model of the loudspeaker to calculate the actual membrane velocity. Finally, the obtained membrane velocity is then multiplied with the pressure results obtained for the unity source.

that the simplified source model neglects the passive admittance term in the loudspeaker representation by implicitly setting $y_{22} = 0$, which corresponds to the assumption of an ideal velocity source with an infinite internal resistance. On the other hand the source term $y_{21}\, U_{\mathrm{inp}}$, which yields the membrane velocity at zero radiation impedance, is substituted by the corresponding membrane velocity obtained for a radiation impedance which is estimated from the considered room sound field using an ideal velocity source. Taking into account that for 'normal' rooms and loudspeakers the radiation impedance of a loudspeaker source is generally very small compared to the summed impedance of the other loudspeaker components (as shown in section 7.3), it can be stated that both simplifying assumptions appear perfectly acceptable for general room acoustic applications.

A.2. A measurement method for the determination of the difference between the free field and pressure sensitivity of a microphone

As discussed in section 6.2 the difference between the free field and pressure sensitivity of a microphone describes the influence of the microphone geometry on the sound pressure at the position of the microphone membrane for a given angle of sound incidence. In particular, the free field sensitivity S_{ff} for incidence angle θ is defined as the ratio of the voltage U_{out} at the microphone output and the sound pressure p_{ff} at the position of the microphone membrane when the microphone is removed from the sound field. The pressure sensitivity S_{pres} on the other hand is defined as the ratio of the voltage U_{out} at the microphone output and the average sound pressure p_{mem} at the microphone membrane when the microphone is present in the sound field. For simplicity we will in the following restrain our considerations to the case of frontal sound incidence on the microphone.

The present section describes a novel precise measurement method for the difference between the free field and pressure sensitivity of a microphone. The measurement method is generally applicable in the frequency range from 100 Hz to 20 kHz and is based on the comparison of anechoic sound field measurements of the considered microphone in the free field and mounted flush into a rigid sphere. In order to extract the influence of the microphone geometry, the measurement results are then compared to analytic calculations of the sound scattering at a rigid sphere, which is exposed to an incident plane sound wave.

A.2.1. Measurement setup

For the measurements in the hemianechoic chamber of ITA at RWTH Aachen University we use a spherical loudspeaker with a single driver which is mounted at approximately 2 m height and at 4 m distance to the microphone membrane. The investigated microphone is a free field equalized *Brüel & Kjær* 1/2" microphone, which is also positioned at approximately 2 m height facing the loudspeaker. The same measurements were also conducted for a free field equalized Schoeps 1/2" microphone, as used in the ITA artificial head. Moreover, the considered microphone is mounted alternately on a thin metal rod with a special unobtrusive microphone clamp and inside a special rigid cube with a diameter of 13 cm, where the microphone membrane is flush with the outer surface of the rigid cube. In order to avoid phase errors in the high frequency range, it is crucial for the measurement setup that the microphone membrane is placed in exactly the same position $\boldsymbol{x}$ for the measurements with and without the rigid cube surrounding the microphone. In order to precisely fix the position of the microphone membrane we have used two laser beams that were crossed at the membrane position. Apart from the exact positioning of the microphone, a very convenient aspect of the presented measurement method is, that the exact type of the used loudspeaker, its precise height above the rigid floor as well as the loudspeaker-microphone distance are not crucial for the method. It is only important that (a) the used loudspeaker does not produce any sharp notches within the considered frequency range, (b) the distance of the loudspeaker and microphone to any reflecting surface (such as the rigid floor in the hemianechoic chamber) is sufficiently large to apply a time window to cancel this reflection and (c) that the loudspeaker-microphone distance is chosen sufficiently large so that the incident sound field on the microphone membrane can in good approximation be modeled as plane wave incidence. The complete measurement setup is shown in figure A.2.

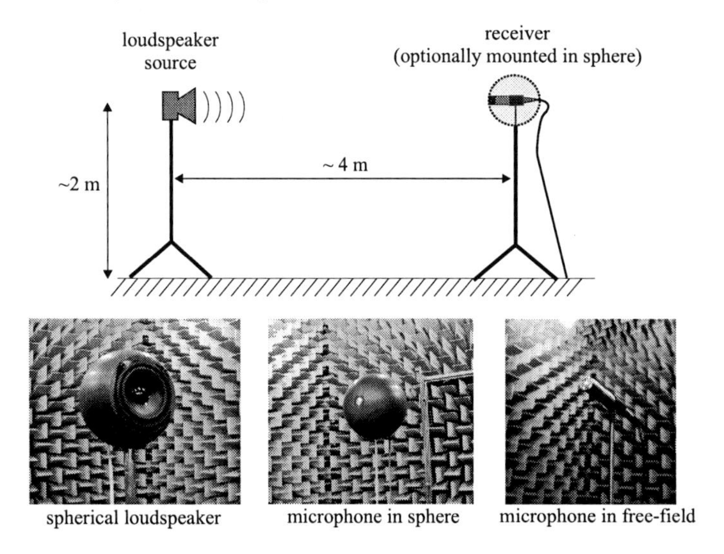

Figure A.2.: Measurement setup for the determination of the ratio of the free field and pressure sensitivity of a microphone.

In order to cancel any influence of the components in the used measurement chain (including the transfer functions of the input and output pre-amps, the loudspeaker and the electrical part of the used measurement microphone) the measurements obtained with and without the rigid sphere in position $\boldsymbol{x}$ are subdivided to give the ratio of the respective microphone output voltages:

$$H_{\text{mess}} = \frac{U_{\text{mic in sphere}}}{U_{\text{mic in ff}}} \tag{A.5}$$

A.2.2. Analytical calculations

The analytical calculations of the scattered sound from a rigid sphere (admittance $G = 0$) for the case of plane wave sound incidence were conducted based on the formulas given in Mechel [2008, pp.189-190] and shall not be repeated here. In particular, the plane wave is propagating in the positive z-direction and is incident on a rigid sphere centered at the origin of the coordinate system with a diameter of 13 cm. The incident and scattered sound pressure (p_{inc}, p_{scat}) are calculated at the position $\boldsymbol{x} = (0, 0, -6.5)$ cm on the outer surface of the sphere. For the comparison with the measured results, the total sound pressure ($p_{\text{inc}}(\boldsymbol{x}) + p_{\text{scat}}(\boldsymbol{x})$) is subdivided by the incident sound pressure at this position, which yields:

$$H_{\text{analytic}} = \frac{p_{\text{inc.}} + p_{\text{scat.}}}{p_{\text{inc.}}} = \frac{p_{\text{analytic on sphere}}}{p_{\text{analytic in ff}}} \tag{A.6}$$

A.2.3. Determination of the ratio of free field and pressure sensitivity of the considered microphone

By using the definitions of the free field and pressure sensitivities, it is possible to write equation A.5 as

$$H_{\text{mess}} = \frac{U_{\text{mic in sphere}}}{U_{\text{mic in ff}}} = \frac{p_{\text{sphere}}}{p_{\text{ff}}} \cdot \frac{S_{\text{pres}}}{S_{\text{ff}}} \tag{A.7}$$

where p_{sphere} is the sound pressure at position $\boldsymbol{x}$ at the microphone membrane when the microphone is mounted inside the sphere and p_{ff} is the sound pressure that would be obtained at position $\boldsymbol{x}$ if both the microphone and sphere were removed from the sound field. Under the assumption that the incident sound field at position $\boldsymbol{x}$ can in good approximation be modeled as a plane wave, the ratio of p_{sphere} and p_{ff} should be the same as the ratio given in equation A.6.

The sought-after ratio of the free field and pressure sensitivity (which is commonly called "free field correction curve") can thus be obtained by a simple subdivision of equation A.6 and A.5:

$$\frac{S_{\text{ff}}}{S_{\text{pres}}} = \frac{H_{\text{analytic}}}{H_{\text{mess}}} \tag{A.8}$$

A.2.4. Verification of method

In order to verify the proposed methodology, we have in a first step run BEM simulations using an ideal point source at 4 m and 100 m distance to the rigid sphere in order to double check the results from the analytical calculations. Despite the different sound field models in the BEM simulations and the analytical calculations, the very good agreement of the obtained results, which are shown in figure A.3, confirm that as expected the chosen source receiver distance of 4 m in the measurement setup is sufficiently large to use a plane wave assumption in the calculations.

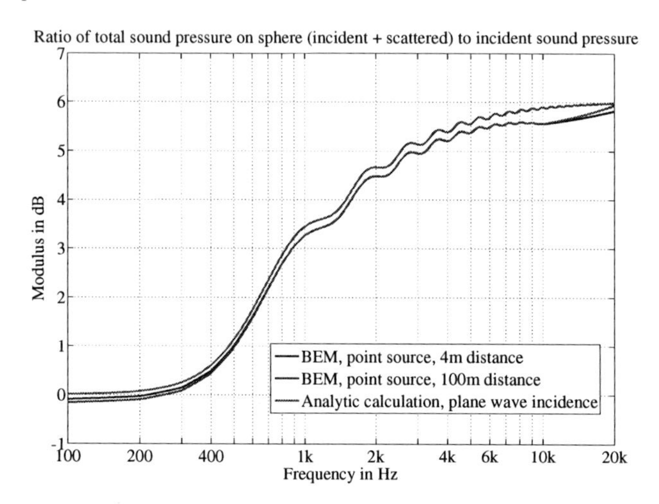

Figure A.3.: Comparison of analytical calculations and BEM simulations of sound scattering at a rigid sphere.

In a next step, the obtained free field correction curves have been compared to the results which were obtained using a calibrated 1/8" *Brüel & Kjær* reference microphone to determine the sound pressure at the membrane of the investigated 1/2" *Brüel & Kjær* microphone for frontal sound incidence in the free field. The measurements were conducted with the microphone setup shown in figure A.4. The 1/8" *Brüel & Kjær* microphone was used since due to its small size this microphone shows almost no influence on the sound field in the considered frequency range. The free field correction curve was thus obtained by a subdivision of the calibrated pressure results from both microphones ($\frac{p_{\text{eighth inch}}}{p_{\text{half inch}}}$). It should however be mentioned that at higher frequencies diffraction effects at the microphone stands and microphone clamps as well as the low sensitivity of the 1/8" microphone might deteriorate the results obtained with this straight-forward method.

Figure A.5 shows a comparison of the results obtained with both methods. Additionally, the figure shows the correction curve as given in a *Brüel & Kjær* technical report for 1/2" microphones[2]. To our disappointment the results of all three methods show rather large differences, which makes it difficult to judge the quality of the new measurement method. By taking the data curve from the *Brüel & Kjær* technical report as a reference, the new proposed method shows a consistent underestimation of the free field correction curve and for frequencies below 10 kHz even worse results than the curve obtained with the 1/8" reference microphone. Since, however, from a theoretical point of view we cannot find any severe flaws of the presented new method further investigations need to be carried out in the future to verify or possibly improve the method.

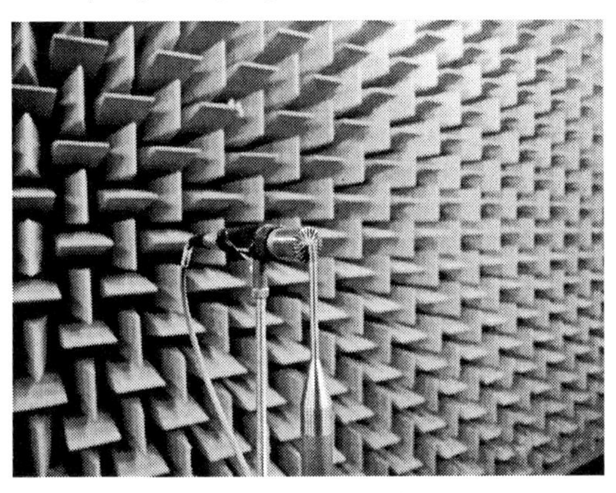

Figure A.4.: Microphone positioning for the determination of the ratio of the free field and pressure sensitivity of a 1/2" *Brüel & Kjær* microphone using a calibrated 1/8" *Brüel & Kjær* reference microphone.

A.3. FE-Coupling of locally reacting two-port network surface elements with fluid domains on both sides

The present section summarizes the FE coupling conditions for passive two-port surface elements which are surrounded by 3D fluid domains on both sides. In the context of room acoustic applications such a passive two-port can be used to efficiently describe a locally reacting layered absorber or partition wall as described in section 5.3. The given equations were originally presented in Aretz [2008a].

In order to couple the airborne sound fields on both sides of a locally reacting wall boundary, the equation for the conservation of momentum (eq.: 3.1 (a)) is used to relate the

[2]The *Brüel & Kjær* correction curve was extracted from the "Technical Documentation, Microphone Handbook For the Falcon Range of Microphone Products", which can be accessed on the *Brüel & Kjær* webpage `http://www.bruelkjaer.de` (last viewed: Jan. 2012)

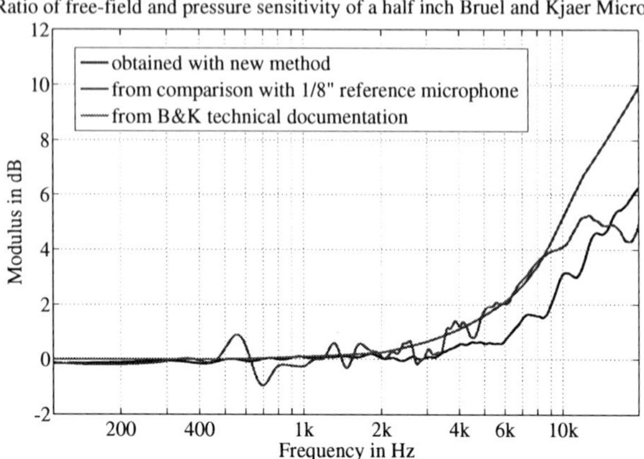

Figure A.5.: Comparison of different methods to determine the free field correction curves for a 1/2" *Brüel & Kjær* microphone.

sound pressures p_1, p_2 on both sides of the locally reacting boundary surface Γ_1, Γ_2 to the respective surface normal velocities $v_{n,1}$, $v_{n,2}$:

$$\nabla p_i \cdot \boldsymbol{n} = \frac{\partial p_i}{\partial \boldsymbol{n}} = -j\omega\rho_0 v_{n,i} \qquad \forall \boldsymbol{x} \in \Gamma_\mathrm{i}, \tag{A.9}$$

where $i = 1, 2$ indicates the side of the absorber. The difference to a simple impedance boundary condition is now, that the pressures and velocities on both sides of the boundary surfaces are interrelated by the sound propagation through the locally reacting boundary layer, which means that the pressure and velocity on one side of the boundary layer depend on the respective quantities on the other side. This relation can be appropriately described by writing the two-port equations for the boundary layer in admittance matrix format:

$$\begin{bmatrix} v_{n,1} \\ v_{n,2} \end{bmatrix} = \begin{bmatrix} y_{11} & y_{12} \\ y_{21} & y_{22} \end{bmatrix} \cdot \begin{bmatrix} p_1 \\ p_2 \end{bmatrix} \tag{A.10}$$

By substituting the surface normal velocities v_{ni} in equation A.9 with the respective terms from equation A.10, a mutual coupling between the sound pressure on both sides is established as a function of the admittance matrix elements. In terms of the integral form of the fluid FEM equations given in equation 4.3) this yields the following coupling conditions for the fluid domains on both sides of the locally reacting boundary wall:

$$\int_{\Omega_1} -\nabla\bar{w}\left(\nabla p + k^2 p\right)\,\mathrm{d}\Omega_1 + \int_{\Omega_2} -\nabla\bar{w}\left(\nabla p + k^2 p\right)\,\mathrm{d}\Omega_2 - ...$$

$$\oint_{\Gamma_1} \bar{w}\; j\omega\rho_0 \left(y_{11}\, p_1 + y_{12}\, p_2\right)\,\mathrm{d}\Gamma_1 - \oint_{\Gamma_2} \bar{w}\; j\omega\rho_0 \left(y_{21}\, p_1 + y_{22}\, p_2\right)\,\mathrm{d}\Gamma_2 = 0, \tag{A.11}$$

where we have omitted all source terms in the fluid domains Ω_1, Ω_2 and all Dirichlet, Neumann or Robin conditions on the non-coupled boundary surfaces of these fluid domains for better readability of the equation. Figure A.6 shows a schematic illustration of the resulting FE system of equations for a very simple example of two coupled rooms and a suitable ordering of the pressure degrees of freedom in the FE mesh. For the FE matrices we use the notation of equation 4.4.

Possible applications of the two-port network surface elements in a room acoustic FE simulation are given in Aretz [2008a] and Aretz [2008b].

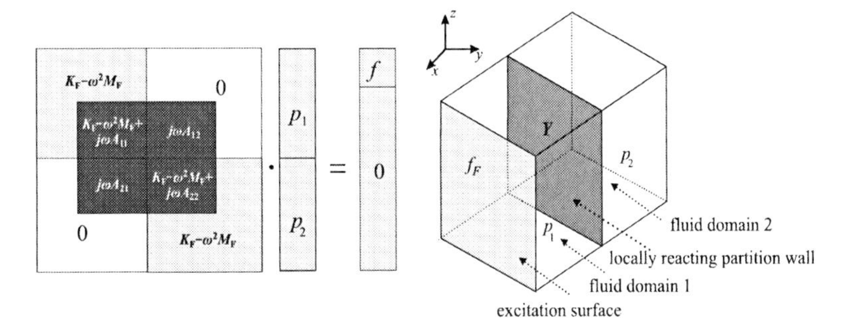

Figure A.6.: Schematic illustration of the FE system of equations for two simple rooms coupled by a locally reacting partition wall (Taken from Aretz [2008a]).
The coupling is modeled based on the admittance matrix parameters of a passive two-port network. The admittance matrix parameters can be deduced from the two-port network model for layered absorbers which is described in section 5.3. Room 1 is excited by a surface velocity on a single room wall. For simplicity all other walls are considered rigid and no further sound sources are present within fluid domain 1 and 2. The notation for the FE matrices is chosen according to equation 4.4. The indices of the damping matrices A_{ij} are chosen according to the indices of the corresponding admittance matrix terms y_{ij}.

Bibliography

J.F. Allard and N. Atalla. *Propagation of sound in porous media: Modeling sound absorbing materials.* John Wiley and Sons, 2nd edition edition, 2009.

J.F. Allard and B. Sieben. Measurements of acoustic impedance in a free field with two microphones and a spectrum analyzer. *The Journal of the Acoustical Society of America*, 77:1617, 1985.

J.F. Allard, Y. Champoux, and J. Nicolas. Pressure variation above a layer of absorbing material and impedance measurement at oblique incidence and low frequencies. *The Journal of the Acoustical Society of America*, 86:766, 1989.

J.F. Allard, W. Lauriks, and C. Verhaegen. The acoustic sound field above a porous layer and the estimation of the acoustic surface impedance from free-field measurements. *The Journal of the Acoustical Society of America*, 91:3057, 1992.

J.B. Allen and D.A. Berkley. Image method for efficiently simulating small-room acoustics. *The Journal of the Acoustical Society of America*, 65(4):943–950, 1979.

J.D. Alvarez and F. Jacobsen. An iterative method for determining the surface impedance of acoustic materials in situ. In *Internoise 2008, Shanghai, China*, 2008.

M. Aretz. Modellierung der Schalltransmission durch lokal reagierende, geschichtete Strukturen mit Hilfe von Vierpol-Flächenelementen für die FEM. In *Fortschritte der Akustik - DAGA, Dresden*, 2008a.

M. Aretz. Different approaches for efficient finite element modelling of absorbers in small rooms. In *2nd ASA-EAA joint conference Acoustics, Paris, France*, 2008b.

M. Aretz. Specification of realistic boundary conditions for the FE simulation of low frequency sound fields in recording studios. *ACTA ACUSTICA united with ACUSTICA*, 95(5):874–882, 2009.

M. Aretz and L. Jauer. Perceptual comparison of measured and simulated sound fields in small rooms. In *Fortschritte der Akustik - DAGA, Berlin*, 2010.

M. Aretz and J. Knutzen. Sound field simulations in a car passenger compartment using combined wave- and ray-based simulation methods. In *EURONOISE, Edinburgh, Scotland*, 2009.

M. Aretz and P. Maier. Application of finite element analysis to the simulation of low frequency sound fields in recording studios. In *VDT Tonmeistertagung, Leipzig*, 2008.

M. Aretz and P. Maier. Simulationsverfahren zur Schallfeld-Auralisation. *FKT - Die Fachzeitschrift für Fernsehen, Film und elekronische Medien*, 4:160–164, 2011.

M. Aretz and M. Vorländer. Sound field simulations in a car passenger compartment using combined finite element and geometrical acoustics simulation methods. In *Aachener Akustik Kolloquium, Aachen*, 2009.

M. Aretz and M. Vorländer. Efficient modelling of absorbing boundaries in room acoustic FE simulations. *ACTA ACUSTICA united with ACUSTICA*, 96(6):1042–1050, 2010.

M. Aretz, P. Dietrich, and G.K. Behler. Comparison of in situ measuring methods for absorption and surface impedances. In *Internoise 2010, Lissabon, Portugal*, 2010a.

M. Aretz, P. Maier, and M. Vorländer. Simulation based auralization of the acoustics in a studio room using a combined wave and ray based approach. In *VDT Tonmeistertagung, Leipzig*, 2010b.

M. Bansal, S. Feistel, and W. Ahnert. First approach to combine particle model algorithms with modal analysis using FEM. In *118th AES Convention, Spain*, 2005.

T.M. Barry. Measurement of the absorption spectrum using correlation/spectral density techniques. *The Journal of the Acoustical Society of America*, 55:1349, 1974.

K.J. Bathe and P. Zimmermann. *Finite-Elemente-Methoden.* Springer, 2001.

G. Benedetto, E. Brosio, and R. Spagnolo. The effect of stationary diffusers in the measurement of sound absorption coefficients in a reverberation room: An experimental study. *Applied Acoustics*, 14(1):49–63, 1981.

L.L. Beranek. *Acoustics.* McGraw-Hill, London, 1959.

J.M. Berman. Behavior of sound in a bounded space. *The Journal of the Acoustical Society of America*, 57:1275, 1975.

M.A. Biot. Theory of propagation of elastic waves in a fluid-saturated porous solid. I. Low-frequency range. *The Journal of the Acoustical Society of America*, 28(2):168–178, 1956.

S.R. Bistafa and J.W. Morrissey. Numerical solutions of the acoustic eigenvalue equation in the rectangular room with arbitrary (uniform) wall impedances. *Journal of sound and vibration*, 263(1):205–218, 2003.

J. Blauert. *Spatial hearing: the psychophysics of human sound localization.* MIT Press, 1997.

R.H. Bolt. Normal frequency spacing statistics. *The Journal of the Acoustical Society of America*, 19:79–90, 1947.

R.H. Bolt and R.W. Roop. Frequency response fluctuations in rooms. *The Journal of the Acoustical Society of America*, 22:280, 1950.

J.S. Bolton and E. Gold. The application of cepstral techniques to the measurement of transfer functions and acoustical reflection coefficients. *Journal of Sound and Vibration*, 93(2):217–233, 1984.

J. Borish. Extension of the image model to arbitrary polyhedra. *The Journal of the Acoustical Society of America*, 75:1827–1836, 1984.

Y. Champoux, J. Nicolas, and J.F. Allard. Measurement of acoustic impedance in a free field at low frequencies. *Journal of Sound and Vibration*, 125(2):313–323, 1988.

A. Cops, J. Vanhaecht, and K. Leppens. Sound absorption in a reverberation room: Causes of discrepancies on measurement results. *Applied Acoustics*, 46(3):215–232, 1995.

T.J. Cox, W.J. Davies, and Y.W. Lam. The sensitivity of listeners to early sound field changes in auditoria. *ACTA ACUSTICA united with ACUSTICA*, 79(1):27–41, 1993.

L. Cremer and H. Müller. *Principles and Applications of Room Acoustics (translated by T.J. Schultz). Vol.I & Vol. II.* Applied Science, London, 1982. Translation with revisions of: Die wissenschaftlichen Grundlagen der Raumakustik. Stuttgart, S. Hirzel, 1948-61.

B.I. Dalenbäck. Engineering principles and techniques in room acoustics prediction. In *Baltic-Nordic Acoustic Meeting, Norway*, 2010.

W.A. Davern. Measurement of low frequency absorption. *Applied Acoustics*, 21(1):1–11, 1987.

H.-E. de Bree. The Microflown: An acoustic particle velocity sensor. *Acoustics Australia*, 31:91–94, 2003.

H.-E. de Bree. *The Microflown E-Book*. Microflown Technologies, Arnhem, 2009. URL http://www.microflown.com/library/books/the-microflown-e-book.html. (Last viewed August, 2011).

H.-E. de Bree, P. Leussink, T. Korthorst, H. Jansen, T. Lammerink, and M. Elwensoek. The microflown: A novel device measuring acoustical flows. *Sensors and Actuators SNA054/1-3*, pages 552–557, 1996.

H. Dekker. Edge effect measurements in a reverberation room. *Journal of Sound and Vibration*, 32(2):199–202, 1974.

M.E. Delany and E.N. Bazley. Acoustical properties of fibrous absorbent materials. *Applied Acoustics*, 3:105–116, 1970.

W. Desmet. *A wave based prediction technique for coupled vibroacoustic analysis*. PhD thesis, KU Leuven, division PMA, 1998.

F. Ehlotzky. *Angewandte Mathematik für Physiker*. Springer, 2007.

V. Esche. Experimentelle Untersuchungen zu Einflussparametern und Größe des Kanteneffektes. *ACUSTICA*, 19:301–312, 1967.

C.F. Eyring. Reverberation time in "dead" rooms. *The Journal of the Acoustical Society of America*, 1:217–241, 1930.

Janina Fels. *From Children to Adults: How Binaural Cues and Ear Canal Impedances Grow.* PhD thesis, Institute of Technical Acoustics, RWTH Aachen University, 2008.

A. Franck. *Finite-Elemente-Methoden, Lösungsalgorithmen und Werkzeuge für die akustische Simulationstechnik.* PhD thesis, Institute of Technical Acoustics, RWTH Aachen University, 2008.

A. Franck and M. Aretz. Wall structure modeling for room acoustic and building acoustic FEM simulations. In *The 19th International Congress on Acoustics, Madrid, Spain*, 2007.

P. Funk. *Variationsrechnung und ihre Anwendung in Physik und Technik.* Grundlehren der mathematischen Wissenschaften in Einzeldarstellungen mit besonderer Berücksichtigung der Anwendungsgebiete. Springer-Verlag, 1970.

S. Ghosh. *Network Theory: Analysis And Synthesis.* Prentice-Hall of India, 2005.

B.M. Gibbs and D.K. Jones. A simple image method for calculating the distribution of sound pressure levels within an enclosure. *ACUSTICA*, 26(1):24–32, 1972.

G.M.L. Gladwell. A variational formulation of damped acousto-structural vibration problems. *Journal of Sound and Vibration*, 4:172–186, 1966.

E. Granier, M. Kleiner, B.-I. Dalenbäck, and P. Svensson. Experimental auralization of car audio installations. *Journal of the Audio Engineering Society*, 44(10):835–849, 1996.

R. Heinz. Binaurale Raumsimulation mit Hilfe eines kombinierten Verfahrens - Getrennte Simulation der geometrischen und diffusen Schallanteile. *ACUSTICA*, 79:207–220, 1993.

R. Heinz. *Entwicklung und Beurteilung von computergestützten Methoden zur binauralen Raumsimulation.* PhD thesis, Institute of Technical Acoustics, RWTH Aachen University, 1994.

C. Hopkins. *Sound Insulation.* Butterworth-Heinemann, 2007.

F. Jacobsen and H.-E. de Bree. A comparison of two different sound intensity measurement principles. *The Journal of the Acoustical Society of America*, 118:1510–1517, 2005.

L. Jauer. Schallfeldsimulation im Fahrzeuginnenraum mittels kombinierter wellen- und strahlenbasierter Simulationsverfahren. Master's thesis, Institute of Technical Acoustics, RWTH Aachen University, 2010.

C.-H. Jeong, J. Brunskog, and F. Jacobsen. Room acoustic transition time based on reflection overlap. *The Journal of the Acoustical Society of America*, 127(5):2733–2736, 2010.

C.H. Jeong, J.G. Ih, and J.H. Rindel. An approximate treatment of reflection coefficient in the phased beam tracing method for the simulation of enclosed sound fields at medium frequencies. *Applied Acoustics*, 69(7):601–613, 2008.

M.E. Kleiner, E. Granier, and P. Svensson. Coupling of low and high frequency models in auralization. In *15th Int. Congress on Acoustic, Trondheim, Norway*, 1995.

K. Knothe and H. Wessels. *Finite Elemente: Eine Einführung für Ingenieure.* Springer, 1999.

J. Knutzen. Untersuchungen zur akustischen Impedanz von typischen Materialien für den Fahrzeuginnenraum. Master's thesis, Institute of Technical Acoustics, RWTH Aachen University, 2008.

T. Komatsu. Improvement of the Delany-Bazley and Miki models for fibrous sound-absorbing materials. *Acoustic Science & Technology*, 29:121–129, 2008.

A. Krokstad, S. Strom, and S. Sorsdal. Calculating the acoustical room response by the use of a ray tracing technique. *Journal of Sound and Vibration*, 8(1):118–125, 1968.

W. Kuhl. Ursachen und Verhinderung systematischer Abweichungen vom " wahren" Absorptionsgrad bei der Absorptionsgradmessung im Hallraum. *ACUSTICA*, 52(4): 197–210, 1983.

H. Kuttruff. On the audibility of phase distortions in rooms and its significance for sound reproduction and digital simulations in room acoustics. *ACUSTICA*, 74:3–7, 1991.

H. Kuttruff. A simple iteration scheme for the computation of decay constants in enclosures with diffusely reflecting boundaries. *The Journal of the Acoustical Society of America*, 98(1):288–293, 1995.

H. Kuttruff. Sound fields in small rooms. In *AES 15th International Conference: Audio, Acoustics & Small Spaces*, 1998.

H. Kuttruff. *Room Acoustics.* Taylor & Francis, 4th edition, 2000.

H. Kuttruff and R. Thiele. Über die Frequenzabhängigkeit des Schalldrucks in Räumen. *ACUSTICA*, 4(2):614, 1954.

K.H. Kuttruff. Sound decay in reverberation chambers with diffusing elements. *The Journal of the Acoustical Society of America*, 69:1716, 1981.

R. Lanoye, G. Vermeir, W. Lauriks, R. Kruse, and V. Mellert. Measuring the free field acoustic impedance and absorption coefficient of sound absorbing materials with a combined particle velocity-pressure sensor. *The Journal of the Acoustical Society of America*, 119(5):2826–2831, 2006.

D. Lee and D. Cabrera. Basic considerations for loudness-based analysis of room impulse responses. *Building Acoustics*, 16(1):31–46, 2009.

D. Lee, D. Cabrera, and W.L. Martens. Equal reverberance contours for synthetic room impulse responses listened to directly: Evaluation of reverberance in terms of loudness decay parameters. *Building Acoustics*, 18(1):189–206, 2011.

E.A. Lehmann and A.M. Johansson. Prediction of energy decay in room impulse responses simulated with an image-source model. *The Journal of the Acoustical Society of America*, 124:269, 2008.

J.L. Levandosky, W.A. Strauss, and S.P. Levandosky. *Partial differential equations: an introduction.* John Wiley & Sons, 2008.

D.Y. Maa. Potential of microperforated panel absorber. *The Journal of the Acoustical Society of America*, 104:2861, 1998.

F.P. Mechel. *Schallabsorber Bd.1. Äussere Schallfelder - Wechselwirkungen.* S. Hirzel Verlag, Stuttgart, 1989.

F.P. Mechel. *Schallabsorber Bd.2. Innere Schallfelder - Strukturen.* S. Hirzel Verlag, Stuttgart, 1995.

F.P. Mechel. *Schallabsorber Bd. 3. Anwendungen.* S. Hirzel Verlag, Stuttgart, 1998.

F.P. Mechel. Improved mirror source method in roomacoustics. *Journal of Sound and Vibration*, 256(5):873–940, 2002.

F.P. Mechel. *Formulas of acoustics.* Springer Reference. Springer, 2008.

Y. Miki. Acoustical properties of porous materials – modifications of Delany-Bazley models. *The Journal of the Acoustical Society of Japan (E)*, 11(1):19–24, 1990.

E. Mommertz. Angle-dependent in-situ measurements of reflection coefficients using a subtraction technique. *Applied Acoustics*, 46(3):251–263, 1995.

P.M.C. Morse and K.U. Ingard. *Theoretical Acoustics.* Princeton University Press, 1986.

G.M. Naylor. Odeon - another hybrid room acoustical model. *Applied Acoustics*, 38: 131–143, 1993.

H. Nelisse and J. Nicolas. Characterization of a diffuse field in a reverberant room. *The Journal of the Acoustical Society of America*, 101:3517, 1997.

C. Nocke. In-situ acoustic impedance measurement using a free-field transfer function method. *Applied Acoustics*, 59(3):253–264, 2000.

R.A. Nöthen. Untersuchung zur Berechnung kombinierter breitbandiger Impulsantworten aus FEM und Raytracing Simulationen. Master's thesis, Institute of Technical Acoustics, RWTH Aachen University, 2008.

J.R. Ohm and H.D. Lüke. *Signalübertragung: Grundlagen der digitalen und analogen Nachrichtenübertragungssysteme.* Springer-Lehrbuch. Springer, 2010.

A.V. Oppenheim, A.S. Willsky, and S.H. Nawab. *Signals and systems*. Prentice-Hall signal processing series. Prentice Hall, 1997.

A.V. Oppenheim, R.W. Schafer, and J.R. Buck. *Discrete-time signal processing*. Prentice-Hall signal processing series. Prentice Hall, 1999.

E.T. Paris. On the reflexion of sound from a porous surface. *Proceedings of the Royal Society of London*, 115:407–419, 1927.

S. Pelzer and M. Vorländer. Frequency-and time-dependent geometry for real-time auralizations. In *Proceedings of 20th International Congress on Acoustics, ICA*, 2010.

S. Pelzer, M. Aretz, and M. Vorländer. Investigation of the effects of temporal fine structure variations on the reproducibility of simulated room impulses responses. In *EAA EuroRegio 2010, Ljubljana, Slovenia*, 2010a.

S. Pelzer, M. Vorländer, and H.-J. Maempel. Room modeling for acoustic simulation and auralization tasks: Resolution of structural detail. In *Fortschritte der Akustik - DAGA, Berlin*, 2010b.

S. Pelzer, M. Aretz, and M. Vorländer. Quality assessment of room acoustic simulation tools by comparing binaural measurements and simulations in an optimized test scenario. In *Forum Acusticum Aalborg, Denmark*, 2011a.

S. Pelzer, D. Schröder, and M. Vorländer. The number of necessary rays in geometrically based simulations using the diffuse rain technique. In *Fortschritte der Akustik - DAGA, Düsseldorf*, 2011b.

M. Pollow, K.-V. Nguyen, O. Warusfel, T. Carpentier, M. Müller-Trapet, M. Vorländer, and M. Noisternig. Calculation of head-related transfer functions for arbitrary field points using spherical harmonics decomposition. *ACTA ACUSTICA united with ACUSTICA*, 2011. accepted for publication, August 2011.

M. Praast. Optimierung und Vergleich unterschiedlicher Verfahren zur Messung der akustischen Impedanz. Master's thesis, Institute of Technical Acoustics, RWTH Aachen University, 2009.

W.C. Sabine. *Collected Papers On Acoustics, Reverberation*. Harvard University Press (Cambridge), 1923.

F. Santon. Numerical prediction of echograms and of the intelligibility of speech in rooms. *The Journal of the Acoustical Society of America*, 59:1399, 1976.

H.A. Schenck. Improved integral formulation for acoustic radiation problems. *The Journal of the Acoustical Society of America*, 44:41–58, 1968.

D. Schröder. *Physically Based Real-time Auralization of Interactive Virtual Environments*. PhD thesis, Institute of Technical Acoustics, RWTH Aachen University, 2011.

D. Schröder and A. Pohl. Real-time hybrid simulation method including edge diffraction. In *EAA Auralization Symposium*, 2009.

D. Schröder, U.P. Svensson, and M. Vorländer. Open measurements of edge diffraction from a noise barrier scale model. *Building Acoustics*, 18(1):47–58, 2011.

M.R. Schroeder. Die statistischen Parameter der Frequenzkurven von grossen Räumen. *ACUSTICA*, 4:594–600, 1954a.

M.R. Schroeder. Eigenfrequenzstatistik und Anregungstatistik in Räumen. *ACUSTICA*, 4(1):456, 1954b.

M.R. Schroeder. Measurement of reverberation time by counting phase coincidences. In *3rd Int. Congress on Acoustics, Germany*, pages 771–775, 1959.

M.R. Schroeder. Natural sounding artificial reverberation. In *13th AES Convention*, 1961.

M.R. Schroeder. Digital simulation of sound transmission in reverberant spaces. *The Journal of the Acoustical Society of America*, 46(2):424–31, 1970.

M.R. Schroeder. The "Schroeder frequency" revisited. *The Journal of the Acoustical Society of America*, 99:3240, 1996.

M.R. Schroeder and B.S. Atal. Computer simulation of sound transmission in rooms. *Proceedings of the IEEE*, 51(3):536–537, 1963.

M.R. Schroeder and K.H. Kuttruff. On frequency response curves in rooms. Comparison of experimental, theoretical, and Monte Carlo results for the average frequency spacing between maxima. *The Journal of the Acoustical Society of America*, 34:76, 1962.

T.J. Schultz. Diffusion in reverberation rooms. *Journal of Sound and Vibration*, 16(1): 17–28, 1971.

H.P. Seraphim. Untersuchungen über die Unterschiedsschwelle exponentiellen Abklingens von Rauschbandimpulsen. *ACUSTICA*, 8(11.52):280–284, 1958.

R.H. Small. Closed-box loudspeaker systems, part I: Analysis. *Journal of the Audio Engineering Society*, 20(10):798–808, 1972.

R.H. Small. Closed-box loudspeaker systems, part II: Synthesis. *Journal of the Audio Engineering Society*, 21(1):11, 1973a.

R.H. Small. Vented-box loudspeaker systems part 1: Small-signal analysis. *Journal of the Audio Engineering Society*, 21(5):363–372, 1973b.

R.H. Small. Vented-box loudspeaker systems, part 2: Large-signal analysis. *Journal of the Audio Engineering Society*, 21(6):438–444, 1973c.

R.H. Small. Vented-box loudspeaker systems, part 3: Synthesis. *Journal of the Audio Engineering Society*, 21(7):549–554, 1973d.

R.H. Small. Vented-box loudspeaker systems, part 4: Appendices. *Journal of the Audio Engineering Society*, 21(8):635–639, 1973e.

G.A. Soulodre and J.S. Bradley. Subjective evaluation of new room acoustic measures. *The Journal of the Acoustical Society of America*, 98:294, 1995.

U.M. Stephenson. *Beugungssimulation ohne Rechenzeitexplosion: die Methode der quantisierten Pyramidenstrahlen; ein neues Berechnungsverfahren für Raumakustik und Lärmimmissionsprognose; Vergleiche, Ansätze, Lösungen.* PhD thesis, Institute of Technical Acoustics, RWTH Aachen University, 2004.

J.W. Strutt. *(Lord Rayleigh) Theory of Sound (two volumes).* New York, Dover Publications, 1877; 2nd Ed. re-issue, 1945.

J.S. Suh and P.A. Nelson. Measurement of transient response of rooms and comparison with geometrical acoustic models. *The Journal of the Acoustical Society of America*, 105:2304, 1999.

J. Summers, K. Takahashi, Y. Shimizu, and T. Yamakawa. Assessing the accuracy of auralizations computed using a hybrid geometrical-acoustics and wave-acoustics method. *The Journal of the Acoustical Society of America*, 115(5):2514–2515, 2004.

U.P. Svenson, R.I. Fred, and J. Vanderkooy. An analytic secondary source model of edge diffraction impulse responses. *The Journal of the Acoustical Society of America*, 106 (5):2331–2344, 1999.

Y. Takahashi, T. Otsuru, and R. Tomiku. In situ measurements of surface impedance and absorption coefficients of porous materials using two microphones and ambient noise. *Applied Acoustics*, 66(7):845–865, 2005.

A.N. Thiele. Loudspeakers in vented boxes: Part 1. *Journal of the Audio Engineering Society*, 19:382–392, 1971a.

A.N. Thiele. Loudspeakers in vented boxes: Part 2. *Journal of the Audio Engineering Society*, 19(6):471–483, 1971b.

L.L. Thompson and P.M. Pinsky. Complex wavenumber fourier analysis of the p-version finite element method. *Computational Mechanics*, 13:255–275, 1994.

J. van Dorp Schuitman and D. De Vries. Determining acoustical parameters using cochlear modeling and auditory masking. *The Journal of the Acoustical Society of America*, 123: 3907, 2008.

J. van Gemmeren. Quantitative Bewertung von Unsicherheitsfaktoren bei der in-situ Messung akustischer Reflexionsfaktoren mit einem Microflown pu-Sensor. Master's thesis, Institute of Technical Acoustics, RWTH Aachen University, 2011.

B. Van Genechten, B. Pluymers, D. Vandepitte, and W. Desmet. A hybrid wave based-modally reduced finite element method for the efficient analysis of low-and mid-frequency car cavity acoustics. *SAE International Journal of Passenger Cars-Mechanical Systems*, 2(1):1494–1504, 2009.

B. Van Hal, W. Desmet, D. Vandepitte, and P. Sas. A coupled finite element-wave based approach for the steady-state dynamic analysis of acoustic systems. *Journal of Computational Acoustics*, 11(2):285–304, 2003.

M. Vorländer. Simulation of the transient and steady-state sound propagation in rooms using a new combined ray-tracing/image-source algorithm. *The Journal of the Acoustical Society of America*, 86:172–178, 1989.

M. Vorländer. Revised relation between the sound power and the average sound pressure level in rooms and consequences for acoustic measurements. *ACUSTICA*, 81:332–343, 1995.

M. Vorländer. International round robin on room acoustical computer simulations. In *Proceedings of the 15th International Congress of Acoustics*, 1995.

M. Vorländer. *Auralization*. Springer-Verlag, Berlin, 2007.

M. Vorländer. Performance of computer simulations for architectural acoustics. In *The 20th International Congress on Acoustics, Sydney, Australia*, 2010.

R.V. Waterhouse. Interference patterns in reverberant sound fields. *The Journal of the Acoustical Society of America*, 27(2):247–258, 1955.

I.B. Witew, G.K. Behler, and M. Vorländer. About just noticeable differences for aspects of spatial impressions in concert halls. *Acoustical Science and Technology*, 26(2):185–192, 2005.

X. Zha, HV Fuchs, C. Nocke, and X. Han. Measurement of an effective absorption coefficient below 100 Hz. *Acoustics Bulletin*, 24:5–12, 1999.

O.C. Zienkiewicz. *The Finite Element Method, 3rd Edition*. McGRAW-HILL Book Company (UK) Limited, 1977.

C. Zwikker and C.W. Kosten. *Sound Absorbing Materials*. Elsevier, Amsterdam, 1949.

Curriculum Vitae

Personal Data

	Marc Aretz
10.10.1980	born in Aachen, Germany

Education

08/1991–06/2000	Secondary School, "Bischöfliches Pius Gymnasium Aachen"
08/1987–06/1991	Primary School, "Gemeinschaftsgrundschule Brühlstrasse Aachen"

Course of Studies

05/2007–04/2012	Ph.D. at the Institute of Technical Acoustics, RWTH Aachen University
04/2007–03/2009	Master's degree in Business and Administration, RWTH Aachen University
08/2004–03/2005	ERASMUS studies of Telecommunications and Information Technology, ENST Paris
10/2001–03/2007	Master's degree in Electrical Engineering, RWTH Aachen University

Employments

since 10/2012	Specialist Acoustics and Vibration (NVH), BMW Group, Munich
05/2007–04/2012	Research Assistant at the Institute of Technical Acoustics, RWTH Aachen University
05/2006–09/2006	Internship at ARUP Acoustics, Cambridge, UK
07/2005–12/2005	Student worker at the Institute of Power Systems and Power Economics, RWTH Aachen University
01/2004–07/2004	Student worker at the Institute of Technical Acoustics, RWTH Aachen University
07/2001–08/2001	Internship at Elotherm AG, Remscheid

Bisher erschienene Bände der Reihe

Aachener Beiträge zur Technischen Akustik

ISSN 1866-3052

1	Malte Kob	Physical Modeling of the Singing Voice ISBN 978-3-89722-997-6 40.50 EUR
2	Martin Klemenz	Die Geräuschqualität bei der Anfahrt elektrischer Schienenfahrzeuge ISBN 978-3-8325-0852-4 40.50 EUR
3	Rainer Thaden	Auralisation in Building Acoustics ISBN 978-3-8325-0895-1 40.50 EUR
4	Michael Makarski	Tools for the Professional Development of Horn Loudspeakers ISBN 978-3-8325-1280-4 40.50 EUR
5	Janina Fels	From Children to Adults: How Binaural Cues and Ear Canal Impedances Grow ISBN 978-3-8325-1855-4 40.50 EUR
6	Tobias Lentz	Binaural Technology for Virtual Reality ISBN 978-3-8325-1935-3 40.50 EUR
7	Christoph Kling	Investigations into damping in building acoustics by use of downscaled models ISBN 978-3-8325-1985-8 37.00 EUR
8	Joao Henrique Diniz Guimaraes	Modelling the dynamic interactions of rolling bearings ISBN 978-3-8325-2010-6 36.50 EUR
9	Andreas Franck	Finite-Elemente-Methoden, Lösungsalgorithmen und Werkzeuge für die akustische Simulationstechnik ISBN 978-3-8325-2313-8 35.50 EUR

10	Sebastian Fingerhuth	Tonalness and consonance of technical sounds ISBN 978-3-8325-2536-1 42.00 EUR
11	Dirk Schröder	Physically Based Real-Time Auralization of Interactive Virtual Environments ISBN 978-3-8325-2458-6 35.00 EUR
12	Marc Aretz	Combined Wave And Ray Based Room Acoustic Simulations Of Small Rooms ISBN 978-3-8325-3242-0 37.00 EUR

Alle erschienenen Bücher können unter der angegebenen ISBN-Nummer direkt online (http://www.logos-verlag.de) oder per Fax (030 - 42 85 10 92) beim Logos Verlag Berlin bestellt werden.